T0220382

RECREATING THE POWER GRID

This book helps power industry executives to systematically navigate the complex technological and organizational changes necessary to recreate power grids.

This is especially pertinent in the current environment characterized by volatility, uncertainty, complexity, and ambiguity conditions. Across the globe, the electric power sector is facing many forces of change as it transitions from a fossil-based system to cleaner sustainable resources. Leaders in the power sector face unprecedented challenges in responding to these changes while continuing to provide safe, reliable, clean, and affordable electricity. Recognizing that historical and existing ways will not work, Jagoron Mukherjee and Marco C. Janssen present a new paradigm for industry leaders to tackle some of the key questions to determine the best path forward: What will the business be like in the future? What technologies will likely prevail? How should my company respond to constant change? How expensive will the transition be? Will customer expectations be met? How fast do we need to change? Drawing on well-known management principles, this book helps industry leaders to provide a methodology to tackle these questions and sharpen their decisions as they embrace innovation, new customer expectations and digitization in their efforts to steer the energy transition.

Taking a holistic problem-solving approach, which addresses the power company as a whole, *Recreating the Power Grid* will be a valuable resource for all professionals working in this quickly evolving field.

JAGORON MUKHERJEE is a seasoned business advisor, speaker, and author who has spent the past twenty-five years working with top electric power companies, industry CXOs, and business executives on a range of business problems. He focuses on helping senior executives cut through the noise and understand the implications of today's fast changing environment. His practical and actionable insights have helped many companies develop strategic plans, technology roadmaps, and realize multi-billion dollar grid modernization programs.

MARCO C. JANSSEN is an experienced advisor, engineer, speaker, and author who has spent the past thirty-two years working with key water and power companies, industry leaders, and business executives around the globe on a wide range of business challenges. He focuses on helping business professionals make sense of technology options, design complex solutions, optimize processes, and understand the implications of today's fast changing environment. His practical and actionable approach has helped many companies design technology roadmaps, improve their operational efficiency, and realize a range of small to large-scale grid modernization programs.

RECREATING THE POWER GRID

NAVIGATING TECHNOLOGICAL AND ORGANIZATIONAL CHANGES

Jagoron Mukherjee and Marco C. Janssen

Routledge
Taylor & Francis Group

LONDON AND NEW YORK

earthscan

from Routledge

First published 2023
by Routledge
4 Park Square, Milton Park, Abingdon, Oxon OX14 4RN

and by Routledge
605 Third Avenue, New York, NY 10158

Routledge is an imprint of the Taylor & Francis Group, an informa business

British Library Cataloguing-in-Publication Data
A catalogue record for this book is available from the British Library

ISBN: 978-1-032-40621-3 (hbk)
ISBN: 978-1-032-40622-0 (pbk)
ISBN: 978-1-003-35399-7 (ebk)

DOI: 10.4324/9781003353997

Typeset in Minion Pro
by codeMantra

CONTENTS

PART 3
PREPARING FOR EXECUTION READINESS **135**

ACKNOWLEDGMENTS

Yogi Berra once said, "It's hard to make predictions, especially about the future." Yet we as humanity have endeavored tirelessly to imagine the future, and the more prudent among us, have invested enormous amounts of time and energy to prepare for it. Such a quest is best done not in isolation, but in the company of supportive colleagues and friends who over the years have critically challenged and provoked us, and in due course sharpened our thinking. This book has been shaped by those discourses that span over fifty years of our collective experience in the power industry. We would be doing a huge disservice if we did not acknowledge those, who mentored, taught, advised, and guided us along the way. The ideas presented in the book have crystallized from the hours of vigorous debates and discussions in conference rooms, field visits, and even pubs and dinner tables.

It will be impossible to list all the names, but a few individuals stand out. For Jagoron, they are earlier mentors and colleagues in Booz & Co.—Joe Van den Berg, Tom Flaherty, Earl Simpkins, Mark Hoffman, Bert Shelton, Owen Ward, Chris Dann, and David Mondrus, all of whom went way beyond the extra mile to give me the opportunities to gain experience, grow, and develop. I also want to acknowledge Robert Carritte, Principal Officer in MPR Associates, who took me under his wing right after I came out of graduate school and taught me the first lessons of client service and consulting more than twenty years back. These individuals have been formidable leaders in the industry and to me personally, a unique gift for which I shall be eternally grateful. For Marco there are several people without whom I would not be where I am today. In particular, I want to acknowledge Maarten Van Riet and Jan Tomassen (may he rest in peace) at NUON (now Alliander) who took me under their wing when I came out of university, gave me their trust and guidance and showed me how to open my mind for the new, with common sense and a dose of humor. I also wish to thank, Roelof Pieters at N.V. SEP (now TenneT) who taught me how to create a proper work life balance and Cees Koreman at N.V. SEP (now TenneT), who educated me on the complex details that make power systems work. I furthermore would like to acknowledge my colleagues and friends at the working groups in IEC TC57, the IEEE, and Cigré who went above and beyond to support me, and last but not least all the colleagues that throughout my career have added to my experience and knowledge.

We were thinking about writing a book on grid modernization for a few years. The complex and systemic nature of the topics that we wanted to share went beyond the discussions that could be rendered in standalone whitepapers and conference presentations, that both of us were collaborating on for a while. In a fortuitous way, the months spent in lockdown during COVID-19, provided the space and opportunity to put together out thoughts into a book.

We extend special thanks to Theodore Kinni, a business writer and editor par excellence, for helping us structure the book. This book also greatly benefited from the wonderful editorial work of Cathy Armer. She was clinical in her approach to ensure that the work was accessible to a wide audience despite the technical and dry nature of the topics.

We thank our publishers to take on this project from us as first-time authors.

And finally, we thank our family—for Jagoron my wife Shveta and son Aadit, and for Marco, my father Marius (may he rest in peace) and mother Emma and my children Alexander and Nannette for supporting us through the long periods of researching, writing, and editing.

This book is written with profound reverence for those who created the electric grid and as a guide to re-create it fit the needs for the next generation. To them, we dedicate this book—our parents, who brought us to where we are, and our children, who will take things forward.

Introduction
The Power Industry Challenge

DOI: 10.4324/9781003353997-1

Across the globe today, leaders in the electric power industry find themselves confronting an "energy transition"— which forces the reordering of our energy system. These industry leaders are at the center of the most profound changes witnessed in the power sector since the beginning of electricity. Yet actions taken to date rarely inspire confidence that enough is being done at the necessary pace to ensure a successful transition. Leaders are criticized for moving slowly, but re-creating the power grid is complex, and poor decisions can lead to results detrimental to their companies and to the well-being of the communities they serve, as well as to their own careers. As they strive to determine the best path forward, power industry leaders are wrestling with the same questions: *What will the business be like in the future? What technologies will likely win? How should my company respond to constant change? How expensive will transitioning be? How do we know our customers will be willing to pay increased costs for energy? How can I attract the right talent? How do we shed old habits and develop new habits and behaviors? How fast do we need to change?*

With several decades of experience in the power industry in various roles—as grid engineers, business operators, financial advisers, and consultants—we found ourselves privileged to be in a position to explore and study these issues in the power industry. It quickly became clear to us that the industry desperately needs a new paradigm, but what does a new paradigm mean for the power business? It means a new set of structures and activities to think about, different trends to observe and scrutinize, new approaches to plan and predict, and a new set of theories and models to make decisions. It is not just a particular function that is affected; rather, it is the entire business that must transform. When we reviewed the existing literature and paradigms, we found a huge body of work discussing techniques and methodologies to deal with general management issues, but nothing mapped to the specific challenges of the power sector. We also found excellent technical material on power system engineering, but these do not make the connections necessary for strategic decision-making and executive actions that cover interrelated policy, business, and technical dimensions. This lacuna prompted us to write *Re-Creating the Power Grid*.

Change is inspiring; it breathes in freshness and evokes the joy of new beginnings. But change is seldom devoid of pain. The change that confronts power industry leaders today is a major institutional change. Institutions, as Nobel Prize-winning economist Douglass North pointed out, consist of formal rules, informal rules, and norms of behavior, may sound obvious, but institutional change is not easy. Institutions in the power industry—the government bodies, regulators, utilities, community groups, and markets—provide stability and predictability. These institutions are not just functions and products of the socioeconomic and political forces of the time; they are formed and shaped by the limits of human capacities, behaviors, norms, and mindsets. They vary from place to place, and when demand for change cuts across all these aspects, it is a tall order for leadership.

This book aims to provide power industry executives with a hands-on guide to lead their organizations through change. Throughout the book we sometimes mention this journey with markers such as "from the old to the new" or "from the 20th-century industry to the 21st century," though we recognize that there is no standard for these markers. This journey is one that all utilities will take, but each enterprise will have its own aspirations and will chart its own course toward its future. As such, this is a journey that calls for the ongoing creation of pathways that ideally start with a course of action, a "point of departure" that leads to the envisioned future. While most seasoned executives are well versed with the issues and challenges facing the power industry, our hope is that this book provides prompts for inquiry that when explored, discussed, and debated will lead to a greater sense of conviction when leaders make their most critical—and often irreversible—decisions.

The changes that the power industry faces today do not originate from within the industry but come from forces outside the industry. *Sustainability* is a major driver for change in our times as climate change is now considered by most people to be the greatest challenge facing humanity. As the globe moves away from carbon and greenhouse gas–producing fossil fuels toward solar and wind as cleaner choices, shifts are taking place not just in the energy domain but in economic, social, and geopolitical domains as well. These shifts have far-reaching consequences; for example, the power balance among countries and economies is disturbed as fossil fuels lose value, the growth in onshore and offshore wind call into question maritime and land rights, and the installation of thousands of rooftop solar photovoltaic systems has created a new class of consumers who are also producers, raising new questions in system balancing, market operations, and business valuation. Beyond the power industry, sustainability choices are also driving the electrification—both in terms of consumption and resources—of other sectors. On the consumption side are transportation with electric vehicles and associated charging infrastructure, as well as heating and cooling; the resources side includes the possibility of using abundant and cheap renewables to produce other sources of energy such as green hydrogen.

Digitalization is another key driver of change today. The growth in information and communication technologies, including artificial intelligence and machine learning, over the past decade, and the digitalization of our society at large is forcing the power industry to adopt new technologies not just for automation and efficiency, but also for providing customers with experiences similar to those that other digitally enabled providers and retailers are offering. In many ways, digitalization is necessary to do business in today's society, but it also has potential to become a point of competitive advantage—by finding new ways to engage customers and launch new business plans, products, and services.

In addition to the two defined challenges of sustainability and digitalization, the overall global environment poses a challenge to power company operations. Recent events such as the disruption of worldwide supply chains during the COVID-19 pandemic, economic stability, migration crises, political instability, and financial disruptions have resulted in a business atmosphere that is highly volatile, uncertain, complex, and ambiguous—a challenge referred to by the abbreviation "VUCA," coined by the US military.

To successfully lead in the VUCA world—which by definition does not provide past data for use to forecast the future—a leader needs a belief system. Decisions based on the past are bound to be wrong both in anchoring the destination point and in determining the path to get there. Successful leaders rely on their belief system to make predictions and then decide and take action accordingly. So where do belief systems come from? A belief system is formed from a collection of rules, hypothesis, and even a theory that can be used to explain reality and in turn be used to make predictions. A good belief system is based on knowledge that comes from a combination of awareness of data and feedback loops that reinforce insights into stronger convictions. Beliefs are not absolute truths, but they must endure; if leaders change beliefs frequently, decisions will not be robust. However, if enough evidence argues against the belief, the belief must be adjusted, and decisions must be changed.

The closest analogy that applies to power industry leaders today is that of the mariner navigating through the seas, charting and changing the ship's course to adapt to ocean conditions that range from periods of turbulence to times of pristine calm. This metaphor came to life in Margaret Heffernan's excellent book *Uncharted: How to Navigate the Future*,[1] and we found "uncharted" to be an apt description for the power industry amid today's uncertainty. Throughout this book, we often use navigational metaphors, though we do not intend to preclude the possibility of industry leaders shaping the environment in which they are operating. Reshaping is possible when there is a spirit of partnership to solve complex matters and when leaders expand their systems of interest (i.e., understanding not just their own organization but also their customers, their regulators, and other stakeholders usually not considered in a unit of analysis). This broad awareness is at the core of systems thinking, and our book draws heavily from the wisdom and teachings of three of the doyens of that discipline, Peter Senge, J.R. Forrester, and Russell Ackoff.

Throughout this book, we return to key themes that contrast to traditional practices:

- There is no one common future state for all; every organization has to envision and create its own future.
- Future events cannot be forecast from the past; rather, they are to be predicted using data and a set of beliefs.

- The gap between the current state and the envisioned future is bound to create tension; leaders should harness the tension to generate ideas and innovative solutions.
- Using a systematic and guided line of inquiry with the right people is a smart way to generate ideas and chart the course of action.
- The course of action is best viewed as a journey in maturity, a stepwise march that is holistic and where everyone plays a part.
- Maturity development is distinct from growth. Development (getting better in capabilities) without growth will matter as much as growth (e.g., empire-building, mergers and acquisitions, and consolidation) without development.

This book is organized into three parts, each containing three chapters. In Part 1, we describe the challenges that face executives regarding the forces of change at play and the critical components necessary to navigate change. Part 2 describes the three new competencies that the power industry has to develop to ensure fitness to navigate change and to be ready to operate and thrive in a changed world that will require agility, flexibility, and adaptability. In Part 3, we provide execution guidance in three areas: business case development, business architecture, and program design and management. Because leaders will rarely involve themselves in the day-to-day details of execution, the focus of these chapters is not on execution details, but on what specifically leaders need to know and what they must do to ensure the future success of their enterprises.

NOTE

1 Margaret Heffernan, *Uncharted: How to Navigate the Future* (New York: Avid Reader Press, 2020).

PART ONE

Forces of Change and Their Impacts

CHAPTER ONE

Forces Impacting the Power Industry

What Are the Forces of Change That Will Require Power Companies to Operate Differently?

DOI: 10.4324/9781003353997-3

For over a century, the electric power industry has been a steady engine driving economic prosperity, enabling many of the benefits of modern life, and enhancing business and industries with unprecedented productivity and efficiency. However, at the start of the 21st century, new forces of change began sweeping and reshaping our lives, forces so profound they rock the very foundation on which the electric power industry was built, challenging fossil fuel use, centralized generation, and government-owned or government-regulated operations. The forces at play span rapid urbanization, decarbonization, renewables growth, decentralization, and digitalization. Often termed "macro trends," the factors shaping the power industry's operational context are coming from outside the industry and are beyond its influence. These trends are often hard to detect in the early stages, yet when they gather full potency, they have disruptive effects on industry operations and economic value drivers. The 21st century has already witnessed dramatic shifts in energy and geopolitics, rapid adoption of renewables and electrification, and increased demand for sustainable living. The industry also has experienced some challenges, as it has swung back and forth between the regulated and unregulated business model. It is clear that these external forces will require the industry to embrace change; the question is simply a matter of when and how.

Change in an industry, even if the need for it is blindingly obvious, is hard and requires leadership, but not just ordinary leadership. Changing the core of a long-standing industry that is strategic and critical for humanity—present and future—under this unfolding environment will require unprecedented and extraordinary leadership. Power company leaders must navigate and negotiate industry changes in the broader turbulent context that preserves the integral role of providing reliable, safe, and affordable power. In addition, leaders need to fulfill the promise of digitalization, sustainability, and changing customer needs, which poses new challenges in an increasingly volatile, uncertain, complex, and ambiguous (VUCA) environment. This chapter provides a grounding to understand the forces of change and global context leaders face as they strive to re-create the power industry for a sustainable future.

FORCES OF CHANGE

Accurate predictions about the future help us make good decisions today. The traditional approach to understanding the future is to isolate external signals of change occurring in political, economic, social, technological, environmental, and legal arenas—six categories captured by the acronym PESTEL. While the PESTEL framework has been useful in the past, the forces of change shaping the future now are dynamic and frequently ambiguous, attributes the PESTEL framework is unable to adequately capture. Because of these forces' complex interconnectedness, they don't always fall neatly into categories, and often the cause and effect among them and between today and the future is not clear.

The macro forces of urbanization and digitalization, and the push for more sustainable and practical solutions to the industry challenges have become mainstream across the globe. Such trends are considered "macro" because they cut across national boundaries, industrial sectors, institutions, and individuals. They may vary in scale, but no one country or industry, let alone one company, however, big and powerful, can influence or shape these forces without collective cooperation. Globalization and interconnected manufacturing and supply chains have altered the mix of energy sources and electricity demand, resulting in shifts in geopolitics and global economics. Events like the 2020 coronavirus pandemic exposed not just the vulnerabilities of these shifts, but also prompted the rethinking of resilience practices, as both the brittleness and the robustness of businesses and countries were evident in how they could adapt and respond and continue to serve their customers and citizens.

In the following sections, we explore six forces of change that are shaping the power industry today:

1. **Decentralization.** Decentralization is the transfer of authority, control, and operations from a centralized entity to entities closer to the point of action. Decentralization is occurring throughout the world today: in media, the advent of personal social media

platforms, such as YouTube has given rise to individual performers generating, distributing, and reaching out to consumers, challenging the large, centralized media outfits that have long controlled content creation and distribution; in the publications sector, big publishing houses are seeing an erosion of knowledge content as self-publishing grows; in finance and trading, the growth of blockchain-based digital currencies or cryptocurrencies, like bitcoin, is becoming channels for currency exchange in place of large and established banks and financial institutions; and in telecommunications, the mobile phone has given billions of individuals access to information and the power that comes from it. In the electric power sector, decentralization refers to producing electricity closer to where it is used.

Distributed energy resources (DERs)—particularly rooftop solar, battery storage, and distributed control systems—now provide a viable alternative to large, centralized generation and vast networks of transmission and distribution (T&D) network infrastructure. Technological advances in DER have led to cost reductions of 90 percent in solar PV cells and batteries, along with breakthroughs in grid edge metering and control systems and artificial intelligence/machine learning-based analytics. Advances in automated grid control ensure faster system balancing that allows for new generation and load profiles. Decentralization is often seen by traditional electric power company players as a problem to solve—for example, leaders look for ways to avoid revenue erosion and loss of customers. But traditional players have an opportunity to transform and align their businesses with decentralization trends.

Decentralization proponents view large regulated centralized entities exercising natural monopoly rights as institutions of an old era and see distributed energy as the way forward. In developed countries, regulatory changes are underway to create rules for commercial markets to enable DERs to transact as part of the optimal resource mix, which includes central generation for the foreseeable future. Advances in grid control ensure system balancing of these new generation sources with changing load profiles. In less-developed countries with weak central power and weak

T&D networks, DER-based microgrids are providing sustainable electrification by overcoming historical financial and political barriers to deployment of large expensive central plants and grid infrastructure. The speed at which these shifts will keep happening is not predictable, and early signals are often too weak to draw attention. But current trends suggest that decentralization drivers in the power industry will gather potency.

2. **Digitalization.** Digitalization or virtualization of physical processes and assets is largely enabled by advances in digital technologies that have transformed our lives over the past two decades. In particular, recent advances in cloud computing, data storage, broadband networks, and communication technologies have given rise to "digital twins" mimicking the physical infrastructure world and providing unprecedented real-time access and interactive control. Virtualization is providing location independence—not just of remote operations of back-office and noncritical tasks performed in virtual servers and storage facilities in the cloud infrastructure, but also of critical tasks. During the 2020 global pandemic, for instance, as remote working became the norm, many critical functions like real-time grid operations, long envisioned as impossible to virtualize, were conducted remotely without traditional physical interventions. Companies made speedy advances and many digitalization plans that originally spanned over two to three years were completed in months.

Advances in robotics, augmented reality, virtual reality, and unmanned vehicles like drones have redefined industries as diverse as telemedicine, construction, and insurance. The new technologies increase efficiency and enhance safety and precision with the aid of advanced algorithms powered by artificial intelligence and machine learning. These advances are affecting the power industry in three fundamental ways: (1) by providing the ability to integrate and aggregate data and work processes to understand, automate, and optimize operations across a broader set of parameters, (2) by adding new capacity with finer control, and (3) by building a brand-new set of services and offerings for customers, creating new markets and revenue streams. These services cover a wide range of digital experiences of interacting and communicating through different voice, text, and web channels. They allow personalized demand management, and they can control systems like electric vehicles and DERs connected to the grid. As new services emerge, they attract new entrants in the industry from startup technology companies to user-experience designers.

In addition, such dramatic technological developments give rise to second- and third-order changes and disruptions in the economy and society. Certain skills become in high demand, while many traditional skills become outmoded. As evident from early movers in Europe and the US, leaders have to navigate complex social and economic inequity challenges that stem from out-of-step virtualization and technological adoption. Many of these challenges are new for industry leaders, and therefore, understanding the full implications on their businesses will be key in deciding the measures leaders take to harness opportunities and preserve the integrity and relevance of their businesses.

3. **Sustainability.** As the world population grows, and with the increase in the level of consumption—from dietary protein intake to electricity and water usage—the global community is increasingly concerned about sustainability. The UN has published sustainability goals,[1] and many companies, including power companies, have embraced them. Green bonds, sustainability financing, and environmental, social, and governance topics are becoming board-level topics. With a finite set of resources and space on the planet, the traditional industrial extractive approach to take–make–dispose is considered unsustainable. The growth in renewable resources is attractive because it directly addresses the sustainability of power generation.

But addressing production-side targets is only one piece of the entire economic cycle. Taking a full-cycle view—that is, addressing what's known as the *circular economy*—means to holistically address sustainability in terms of climate, biodiversity, waste, and pollution. It is not just about the environment; it is about achieving a systemic balance between the environment, economy, and social issues. Sustainability issues even reach into matters like child labor, money laundering, and human rights, traditionally outside the purview of industry leaders. Leaders need to use a wider lens to monitor and track these developments and how they affect their own companies. For the power industry, it means thinking beyond solar and wind as renewable resources and as replacements for fossil fuels or nuclear power to understanding the sustainability of entire supply chains, finding counterbalances to the finiteness of the metals and minerals that are used for solar panels, and considering the use of water and other natural impacts on flora and fauna through the production and operations cycle. Additional sustainability components are addressing disposal and minimizing waste. Reprocessing and reusing the waste to minimize the residue on landfills, oceans, and the planet completes the circular economy.

Along these lines, innovations are already underway regarding repurposing used solar panels, for example, and many investors are vocalizing the need to factor after-life disposal costs in their business cases. In many parts of the world, we already see this trend as governments and businesses shift from plastic to biodegradable materials. Advances in battery storage and in cheaper renewable power are leading toward the mass electrification of various sectors. The electric vehicle industry is expected to phase out gasoline-powered vehicles in the near future. Other industries—for example, cement, paper and pulp, mining, and airlines—will take longer and will likely need breakthrough innovation, but the motivation for sustainability is striking, with research investments in biofuels, hydrogen, and other advanced fuels.

4. **Climate change.** Climate change is one of the greatest challenges facing humanity, with deep repercussions for today's society. The climate problem is complex and multidimensional, and it transcends geopolitical boundaries.

   As cause and effect are not always observable, countries and companies must strike a balance between longer-term impacts and near-term investments and between individual versus collective action. The air and the climate are public goods,[2] but there is no international governing body to manage their quality, and driving consensus on measures to reverse human-induced climate change among all the nations is an arduous task. The Paris Climate Accords signed in 2015 by 196 countries was a step toward a global consensus, but as a voluntary agreement, it is structurally weak, and countries can deprioritize their engagement. Recent Intergovernmental Panel on Climate Change studies make dire predictions unless humanity takes drastic actions to reduce greenhouse gases. Sharing technology and providing economic assistance are contentious issues as different stakeholders weigh their own incentives and goals, which can conflict with climate priorities. Poorer countries often rely on cheaper fossil fuels, and these countries are typically not well positioned to invest in measures because of capital and technological limitations.

Climate change also creates economic and social dislocations that severely impact less-developed countries, whose economies are agrarian and more dependent on weather. As extreme weather events have increased, with more wildfire events, droughts, superstorms, excessive rainfalls, and temperature fluctuations, leading eventually to mass migration to cities or to other countries, and often resulting in geopolitical tensions. We are already witnessing the effects of climate change on produce with shifts in harvesting seasons. Wildfires are increasing in places like California in the US, as well as in Australia and Turkey, a consequence of increased planetary warming, and often sparked by power grid assets, pushed to operate outside their design limits. Increased flooding of coastal areas is observed as sea levels rise from the melting ice at the globe's poles.

Climate change has both direct and indirect impact on the power industry. On the one hand, it pushes for a transition from fossil fuels toward increased electrification, and on the other hand, the industry has to account for the readjustment of demographics, industrial shifts, and consumption patterns—all happening simultaneously and making predictions extremely difficult.

5. **Urbanization.** According to UN estimates, by 2050, about 70 percent of the global population of 10 billion people is expected to live in urban areas. This rural-to-urban shift is already being seen in Asia, Africa, and Latin America, where economic opportunities in the cities relative to those in rural areas keep growing. Rapid population growth creates enormous demands on city infrastructure and for basic needs like electricity and water. Even developed countries, which have seen either slow or stagnant electricity demand growth compared with GDP growth; have experienced population migration. According to recent US census, several southwestern cities in the US added population, while northeastern cities saw a population decline. As cities are stressed, there is a growing trend in deploying digital technology solutions as part of "smart city" sustainability initiatives. These "smart" solutions for cities are diverse and complex multistakeholder undertakings.

The power industry is a critical player in citizen services, but successful citywide solutions require new models for collaboration and working with other city utilities and service entities both at technical and policy levels. Every urban situation will have its own set of challenges and problems to solve; for example, the challenges of Amsterdam are quite different from those of Dhaka, and the governance of Chicago is different from that of Dubai. To chart their courses effectively, power industry leaders will need to understand the big picture and the specific context of a city's needs, and they will need to connect the policy, economic, social, and technological dots.

6. **Customer expectations.** Customer expectations—driven by what customers need and what they want, which is in turn shaped by what other companies are offering, are shifting. As digital platforms have removed communication and congregation barriers, customers have unlocked their ability to express opinions on the products and services they receive from their companies to a wider audience. Customers have a voice and a platform, and many take an activist posture. This development is a new consideration for power companies, and sadly, it is one many have come to realize only when caught on the wrong foot.

In the traditional regulated environment, due to a structural asymmetry, public activism against power companies was infrequent. The cost versus benefits for an individual customer to organize a campaign were disproportionately skewed against the customer. While digital technology provides customers with a new voice, it also provides companies the tools to meet new expectations. Technology companies can observe patterns of customer behavior, develop insights, and take proactive measures to mitigate public relations risks and to create new business opportunities. Companies like Amazon have "customer obsession" central to their business principles,[3] resulting in tremendous value creation. Uber and Airbnb have become ubiquitous poster children of digital platform-based business models providing a complete, low-friction user experience. Netflix gathers customer data to understand the movies and shows its customer what they might like. Social media companies are constantly updating their business models to provide users with innovative and often complex multisided market[4] offerings and services. Customer attention has become the new currency for value creation. And for attention, personalization—only possible with data and computational analytics—has become a key to success. Digitalization provides a new realm because it not only offers customer awareness and insights but also allows constructs such as modularity, over-the-air software updates, and automation to react to and address customer needs.

In industry after industry, traditional business models have collapsed, from newspapers that moved from printed delivery to a web- and mobile-centric model with new subscription options. The shelf life for new business models has also shortened. As businesses are investing hundreds of billions of dollars in cloud computing, artificial intelligence, and machine learning, power companies will have to align with the evolving needs and desires of their customers. Power industry leaders not only have to observe and learn from what customers are asking of them, but also they must understand how the needs and wants are shaped. For a better understanding of customer expectations, the power industry needs to look to other industries such as banks, grocery stores, education centers, and retailers.

IMPLICATIONS FOR UTILITY INDUSTRY LEADERS

Given that these changes are rooted outside the industry and are pervasive across many industrial sectors, power industry leaders' sphere of direct control is limited to anticipating the challenges faced and steering the industry with the tide or against it. Hence, leaders need to be vigilant regarding activities and events at the edge of the traditional power industry boundary, and they even need to spot for emerging trends in other industries that very soon may become relevant and even table stakes in their own businesses. Finally, the systemic and dynamic nature of the changes, described earlier, creates a VUCA environment within which to make strategic investment and management decisions; leaders should be prepared that their solutions will fall in gray areas. Here, we describe in greater detail the VUCA environment as it pertains to the power industry.

VOLATILITY

Large and drastic shifts in markets and technology result in increased volatility in prices and output. Such changes are common in deregulated markets, and companies hedge ensuing risks with sophisticated financing structures. But increased volatility is endemic to today's power industry. The growth in renewables and the resulting shift in the energy mix, policy changes, and increased adoption of modern technology have introduced and increased volatility across the value chain. The use of subsidies in the form of tax credits and feed-in tariffs to promote renewables over cheaper fossil power resources created price volatility, and as renewables prices reached parity in many places, the interplay of policy changes, new grid operating patterns, and technology interventions added operational volatility to the system. Intermittency in the power system results in sharp changes in the production and call for countermeasures in either curtailment or fast ramp-ups. The interconnectedness of supply chains—currently prone to disruption and shortages—has added to volatility. Finally, cyberattacks can change economics and timelines considerably.

UNCERTAINTY

The future has always been uncertain, but it is even more so now. With changes occurring in so many interconnected areas simultaneously, an event in one corner of the world can ripple through countries and industry sectors quickly, causing disproportionate impacts. Given large networks of global supply chains and platforms of connected systems, it is impossible to predict the course of the future. We simply don't know how certain events and developments may impact our society and our economies. The global 2020 coronavirus pandemic—with the subsequent shutdown of international travel and large parts of the world's economic ecosystem—is a recent example. Looking back a decade or two, the rapid reduction of renewables costs, the speed of the electrification of formerly fossil-based transportation and industries, the power and energy density, and the sustained viability and scalability of battery technologies are all creating unknowns and driving radical shifts that could not have been predicted. Moreover, technology lifecycles have become shorter, and radical improvements can come from practically anywhere. Non-traditional startup companies from any part of the world—and from outside the four walls of the industry—are increasingly becoming a promising source of innovation compared to traditional industrials.

The fragmented nature of these developments and the enormous burden to scan and track trends add to uncertainty. Yet businesses and investors look for certainty and attach risk-adjusted price premiums to increased levels of uncertainty. While science and economics strive to reduce uncertainties, most techniques are based on historical data—despite the fact that today's environment is such, that models and known methods are no longer sufficient to remove the uncertainties. In fact, the traditional approach—to know enough about the outcome to make a decision—is wishful thinking. Uncertainty is the new norm under which companies have to forecast their demand, allocate their capital investments, hedge their risks, design and manage their programs, and develop talent and skills.

COMPLEXITY

Complexity arises from the connectivity and interdependence of the different components of the forces of change. As the World Economic Forum highlighted:

> The complexity of energy transition results from the diverse components within the system itself, as well as their interdependencies with components outside the energy sector. The energy system's boundaries include different fuel sources, extraction and conversion processes, and infrastructure, workers, investors, innovators, and different end-use sectors. Beyond the boundaries, energy is a commodity traded between countries and a key component of public policy within them. The volatility of energy markets and trade flows influences countries' fiscal and monetary policies. The energy system also enables economic growth by fueling industrial activity, providing employment and creating national income through exports. Universal access to energy is important to alleviating poverty and improving outcomes on social objectives, such as education, health and gender equality.[5]

Complexity in the power industry and in the broader energy sector is already evident in the early experiences with transition. Countries that took aggressive policies in supporting renewables through subsidies, either in the form of feed-in tariffs or tax incentives, had to manage a new set of imbalances as penetration of decentralized rooftop solar PV cells increased, and economic pressures were mounting on traditional utilities. Germany saw its major utilities lose huge market capital at a rapid rate that led to a painful[6] restructuring of the German energy companies E.ON and RWE. In the US, DER growth moved at a rate that gave rise to free-riding issues with legacy net metering policies, leading to a series of contentious on- and off-the-court battles over new policies on grid access fees and demand charges. Grid planning and resource adequacy issues arose, causing blackouts and exposing reliability bottlenecks. Technical issues resulted in economic challenges that led to social challenges. The nature of transition exposed inequities in society, where the affluent was able to take advantage of new developments, sometimes at the cost of the marginalized. In California, which has a tiered tariff structure, the communities in higher tariff brackets could take advantage of switching to self-generation, placing a higher burden of the utility revenue requirement recovery on those who could not afford rooftop PV systems. Where renewables reached economic parity, retirements of fossil and nuclear power plants happened at a pace faster than the replacement with dispatchable resources and storage deployment, resulting in capacity shortages and stranded asset problems.

Extreme weather events attributable to climate change have exacerbated these issues and have brought into question the legacy planning philosophy. One-in-ten-year events used for reserve margin and loss-of-load events planning are happening more frequently, creating unprecedented electricity shortages. Repeated shortages have led to economic and political angst and consumer anger in different parts of the world. All these situations highlight the complexity of the operating environment, in which the traditional approach of treating issues one at a time in siloed piecemeal parts is grossly inadequate. With the systemic and interconnected nature of the issues managed by various stakeholders, many positioned outside industry boundaries, leaders find that they must address complex problems in an increasingly interconnected world.

AMBIGUITY

A complex and uncertain environment introduces ambiguity in decision-making. Cause-and-effect links are not always observable or fully understood, and any intervention has to consider multiple possibilities and be evaluated from different perspectives. For example, when Germany decided to exit its nuclear fleet and initiated its energy transition program (*Energiewende*), one result was a tremendous growth in renewables. At the same time, the nuclear capacity gaps could not be fully addressed by renewables, and those gaps were filled by increased reliance on Russian gas imports, and an increase in lignite generation, a high emitter of carbon. Thus, net emissions increased despite record growth in renewable generation. Further, growth in renewables along with the retirement of fossil power plants in other places has interplayed in ways that resulted in capacity constraints in operations, operational and reliability issues, lowered energy security, and broader economic and social problems created by the loss of high-paying fossil plant jobs, price cannibalization, and devaluation of many parts of the industry value chain. Along with advances in the application of digital technology and data-driven decision-making, safety and performance have improved, but digitalization also creates major shifts in the workforce resulting in job losses for those who cannot upskill or relocate.

Customer engagement is also changing. Companies are expected to make the necessary investments and elevate customer engagement and communication, but customer experience can be highly subjective. Historically, reliable service and efficient account management were sufficient to maintain customer satisfaction levels. But when customer needs and wants are changing and vary widely, companies struggle to determine which products or processes will resonate with customers. Furthermore, ambiguity often causes conflict within organizations, so many leaders try not to deal with it and seek out the status quo. In the current environment, dealing with ambiguity will be essential and will require skills not just for individual leaders but for all participants in the power industry.

SUMMARY

This chapter illustrates the need for leaders to expand their range of knowledge and acknowledge the forces of change impacting the power industry. Understanding specific change drivers is critical to successfully navigate and steer the power industry to a sustainable future. Never since its inception or through its maturity, covering more than one hundred years, has the industry faced such radical changes in the external environment. Traditional models and patterns for creating value in the power industry are proving inadequate in the face of such changes; for example, fossil-based generation is not sustainable nor is climate-friendly and nuclear is not deemed safe or economical. Customers are calling for deeper and more interactive engagement. Digital technology-enabled virtualization requires assets to communicate and processes to be automated, presenting radical operational improvement opportunities, and creating new revenue opportunities. Careful selection of digital solutions is also at the forefront for tackling the growing urbanization trend that puts cities under stress to provide electricity, water, and other services in a sustainable and cost-effective manner. Digitalization brings its own set of new challenges, including cybersecurity issues. Increased reliance on electrification also increases vulnerabilities in supply chains and to extreme climate events that have become more pronounced and even spiral into geopolitical issues. These unprecedented changes are neatly captured by the US military–coined acronym VUCA. Leaders have a responsibility to navigate their organization through the uncharted waters of this VUCA world. The future is here, and a management system is needed to meet new demands along with the long-standing mandate to deliver safe, reliable, sustainable, and affordable power. With this contextual backdrop, our next chapter discusses specific leadership challenges.

NOTES

1 UN Sustainable Goals – https://sdgs.un.org/goals.
2 "Public goods" are goods that are important and of benefit to all members of society and users cannot be barred from accessing or using, even if them fail to pay. Examples of public goods are national security, street lighting.

3 See "Customer Obsession" under "Leadership Principles," Amazon, https://www.amazon.jobs/en/principles.

4 Multisided markets are enabled by an intermediary economic platform that has multiple, distinct user groups that provide each other with network benefits. The organization that creates value primarily by enabling direct interactions between two (or more) distinct types of affiliated customers is called a "multisided market platform." An example is Google, which provides search benefits to users and also offers a platform for advertisers who generate Google's revenue by targeting the user base.

5 World Economic Forum, "The Scale and Complexity of Energy Transition."

6 More than 60 percent of market value was lost in 2015.

CHAPTER TWO

The Leadership Problem

How to Turn the Long-Standing Power Industry into a Startup?

DOI: 10.4324/9781003353997-4

A few years ago, the head of the newly created digital business unit of a large industrial conglomerate shared with us his struggle to get equity investors to value the company as a software company. He drew on the example of Tesla, an automaker that was valued at price to earnings (P/E) multiples of software companies, not automobile companies (whose valuation was stagnant). When we asked about the leader's goals and how he anticipated reaching them, he acknowledged that he did not have all the answers, but he knew investing in marketing campaigns that branded the company as a software company was necessary. During the working sessions with the company's other business units and line functions, gaps in alignment and collaboration were glaringly obvious. The marketing units flushed with funds were focused on storytelling and messaging to customers and investors the merits of the new digital business. But the engineering and product management units had no guidance, direction, or focus other than some funds to support marketing and sales efforts. After a few product acquisitions, several internal initiatives were abruptly stopped, leaving the engineers and developers confused and demoralized. The new digital product never got beyond a few pilots used by early adopters. There was no plan for the legacy core business units responsible for current revenues and margins to absorb the new products. Leadership was single-mindedly focused on a valuation uplift without a vision for product development, scaling, and growth. Customers and investors went along for a couple of years, but never bought on to the vision of the new business. After millions of dollars spent, the digital business unit was downsized and practically disbanded.

Such stories are not uncommon. There are very few success stories in which matured companies seeded and scaled a new business with new products and business models, funded by their legacy business' retained earnings but demanded a fundamentally different operating structure. We've discussed challenges facing the power industry leaders in the VUCA environment. The way the business must be transformed requires leaders to thoughtfully frame and formulate the strategic directives that fit their organizations. In the example above, one may wonder whether framing the aspiration as a Wall Street valuation uplift was the right choice. Would a more universal vision of the future have been more powerful? In the face of uncertainties, leaders must decide which approach will work best. Leaders must connect and deal with uncertainty in two distinct tracks—uncertainty in the environment, which we discussed in the previous chapter, and uncertainty regarding what to do, discussed in this chapter.

FRAMING THE PROBLEM: GETTING THE DIAGNOSIS RIGHT

Herbert Simon, a Nobel Prize-winning pioneer in the art of problem solving, developed a body of work that showed that the difference between expert and novice problem solvers is in how experts represent the problem.[1] As skilled problem solvers, leaders know that the likelihood of a good solution lies in good framing of the problem, which begins with the question, "What problem are you solving?" This approach may seem obvious, but with the myriad of issues flooding leaders every day, diagnosing, framing, and communicating the right problem for the organization's focus can be a challenging task.

When we ask industry leaders about their most pressing problem, we often hear of stagnant revenues and of demand not growing while costs are. Growth in DER and in private generation creates increased threats to future revenues of legacy companies and underscores the need to upgrade infrastructure. But regulated businesses exercise caution and do not like taking undue risks or betting on uncertain technology. With rapid changes in technology, and as it often happens, new technology emerges before the current technology is fully analyzed, vetted, de-risked, and deployed. Such situations present an uncertainty conundrum. Moreover, recent technology has increased the volume of data collected, but it is not always clear a-priori how to use and monetize the data for value creation. The vulnerability of cyberattacks has intensified often requiring companies to make wholesale re-design of business processes. Indeed, companies need to move quickly in this dynamic environment, yet legacy structures are often bureaucratic and slow to change. Many executives believe that the only way to grow is inorganically, through mergers and acquisitions, because legacy structures inhibit organic innovation and growth. The markets and investors reward certainty over unbounded risk-taking—so even though the need for innovation is obvious, the real barrier is the power of the status quo.

Despite the wide variety of issues that industry leaders have to deal with on a daily basis, some common themes underpin these situations. Key is that the pace of change demands a much faster and robust response from the industry. But even if matured businesses wake up to this need for speed, success stories are utterly few. Change, as in renewal, is not without pain, and it takes tremendous leadership and a dedicated organization to pursue the renewal process. Moreover, the change confronting the power industry is a fundamental change, often termed as *institutional change*, which includes not just structural aspects such as assets, networks, facilities, systems, processes, and organizations, but also behavior, norms, and mindsets. As one executive of a large industrial entity shared, "Even our fastest day is not fast enough. We need to let go of many of our old ideas. We must start acting like a startup, while our existing business is still running." This executive has identified the tension between balancing the daily demands of the power industry, which depends on a legacy architecture, and the need to turnaround and build an agile and flexible organization that can pivot to respond to a dynamic environment. In short, a complete overhaul of the business is needed, while efficiently performing the current functions of a critical infrastructure.

TURNAROUND TO ABSORB STARTUP ATTRIBUTES

When we imagine a startup business, we envision a company with a limited operating history that is in a discovery phase of a product-market fit despite looming uncertainty of success or failure. We imagine the company facing a series of fits and starts, trials and tribulations, yet the founders and employees finding ways to seed ideas and become competitive in the marketplace. Although the term *startup* became popular in Silicon Valley and is used for venture-backed firms, we use it here to describe a new business initiative that is seeded to scale and is future growth-focused and setup to deliver a high risk-to-reward outcome. In certain situations, the business initiative may even create or shape a market, in which case the scaled startup occupies a de-facto market leadership position. This approach is what the power industry needs. We have identified six success factors of the startup model relevant to the power industry:

1. **Novelty.** Startups are associated with creating something that has not been experienced before. For our purpose, *startup* is a byword for fostering a creative environment. But creativity is not about the theatrics of flip charts filled with colorful sticky notes. Creativity is about exploring and discovering new ways of solving problems. It is about developing new concepts and new units of analysis, finding evidence and facts, and creating measures to manage complex multidisciplinary industry challenges.

2. **Think big, start small, iterate, and then scale.** Startups begin their journey with no solid customer base, with a market that is immature and does not yet exist, and with a product or service that is not fully developed. Despite uncertainty, startup aspirations are big, and they take concrete actions, test their ideas, learn from failure, iterate, and ultimately build and scale.

3. **Staged, gated venture funding.** Startups secure funding in increments linked to the incremental reduction of risks as the new venture progresses from early stage to scaling. Such a funding mechanism is an extremely useful approach, given the risk-averse disposition of most power companies. Similar to how funding allocation is often linked to specific milestones such as developing a working prototype, finding a first customer, or achieving commercial breakeven, power companies can adopt similar gates to allocate funding based on reductions in risks on their path to self-sustaining ongoing investments with the value created.

4. **Market- and customer-backed.** When the customer base (either internal or external) is established, as is the case in the power industry, the startup mindset must work backward from the market and customer pain points to develop products and services. This approach keeps a sharp focus on the business rather than on developing a solution and then looking for a problem, which many companies fall prey to, especially with the temptations of the latest technology.

5. **Minimum viable problem over minimum viable product.** A changing environment presents a continuous stream of challenges to overcome, maneuver through, and negotiate. Leaders need to "pick their battles," which calls for judicious selection of which problem areas to explore, learn, and commit resources. Such decisions must be fast and cost-effective to manage the volumes of hypotheses that are to be tested—an approach different from traditional problem solving that involves testing three to five hypotheses over several years. To adopt agility, leaders must discipline themselves to address the minimum viable problem that is worth working on—akin to the minimum viable product that many startups and technology companies also limit themselves to. Identifying what not to work on is as important as what to do.

6. **Experimentation mindset.** A matured company aiming to encourage agility akin to a startup must formalize a structure that provides permission to challenge the status quo and embrace change. This management structure includes hard assets, policies, processes, and governance with rules of engagement for scanning, investing, and seeding new ideas and bring them to fruition at scale. To further incentivize experimentation, the management structure needs to be complemented by a mindset that looks around the corners with a beginner's mind and embraces failure as a learning process.

FOCUS ON THE WEAKEST LINK

Many power companies and utilities have realized the importance of the startup model and have set up dedicated incubators and ventures, often located in innovation hubs such as the San Francisco Bay Area, Berlin, and Bangalore. Their office design, talent acquisition and retention model, incentive structures, and culture aim to foster the environment of the most innovative technology companies. But the aspect that often proves to be the Achilles' heel of their efforts is *transition*. How to adopt a separate venture with its agile innovations into the fold of the legacy company? How can the old company embody the innovation and associated changes at scale and become part of core operations?

Leaders who sponsor seeding and growing of new ideas to scale must act like skilled gardeners, aware of the fragility of their undertakings and the likelihood of early mortality. A new venture is particularly vulnerable when it moves into the folds of the matured and long-established company. Most companies do not pay much attention to this transition until they face resistance. Too often, leaders assume that because the venture is good, it will be embraced, and transition will be easy. The few companies that are consistently successful with transition conduct a thorough diagnosis of the points of vulnerability, understanding that the weakest links in nurturing innovation and absorbing it to the core business are both structural and cultural. In this case, leaders are better off leading with the targeted structure and a transition strategy, because without adequate processes, performance incentives linked to adoption, and roles and responsibilities redefinitions, the power of legacy culture is likely to eat the new initiatives for breakfast.[2]

THE LEGACY PROBLEM

To successfully adopt the best of the startup model and become an agile and nimble organization, leaders need to make a number of changes to legacy practices. These changes are not just about hard-to-change processes and systems but also extend to the interactions with partner ecosystems, collaborators, and customers. The following are key areas where leaders must pay particular attention:

1. **Business continuity, efficiency, and growth.** Power company leaders have to undertake the transition journey while fulfilling the mandates of current business as well as meeting growth expectations in the current market environment. The continuous growth model has been fueled mostly by cost reductions resulting from productivity improvements. Improvement levers include the automation of routine tasks using computer-based solutions, integration of processes and systems, consolidation, and outsourcing to third parties. Particularly in developed countries, as electricity consumption and top-line growth have slowed, focus on cost reductions has been the dominant lever of growth. For most power companies, as discussed earlier, creating new growth opportunities organically from within the organization will require a different kind of leadership that is able to balance ongoing priorities with the ability to navigate changes.

2. **Fighting matured company inertia.** Many decades of successful operations have created strong institutional behaviors that go beyond formal structures and policies. Informal norms and practices become deep-rooted and impossible to eradicate overnight. Leaders will have to find ways to manage a systemic effort to overcome and break down inertia to become more adaptive, flexible, and nimble. While the need to overcome inertia is crucial, care is required: rapid policy changes have at times caused severe, unintended effects ranging from drops in reliability, service levels to corporate bankruptcy.

3. **Overcoming deterministic planning.** Companies must become agile and nimble to successfully cope—and thrive—under uncertain conditions. The traditional enterprise asset planning in the power industry often tends to use forty- to sixty-year cycles and is not used to asset renewal planning of much shorter five- to ten-year time periods. Between 2010 and 2020, when solar PV and battery costs dropped by more than 90 percent, many companies were caught off guard and could not pivot quickly to reallocate funds from legacy R&D programs (e.g., advanced gas turbines). Many kept betting on fossil fuels and did not reallocate their resources

from legacy programs. When companies are set in their ways, the appetite and incentives for experimentation are absent and even considered imprudent. Recognizing that today's technology-driven changes do not offer a lifecycle and a line of sight of forty to sixty years, leaders must inculcate a discovery-based planning approach that is exploratory and flexible. Such an approach does not mean exploration for exploration's sake, but rather it means using a clear direction to enable rapid experimentation of initiatives; to verify fit with the changing environment. Plans must allow abandoning nonworkable projects quickly and reallocating resources to prevent waste.

4. **Regulatory hurdles.** Regulatory mandates are designed to ensure customers are not squeezed for profligate or irresponsible spending by power companies. In its current design, regulation discourages companies from taking out-of-the-ordinary risks. With the growing risks and uncertainty of the operating environment, the bias toward the status quo not only stifles innovation but also increases the risks of omission (i.e., not doing what is necessary and missing the opportunity). Leaders have limited agency in changing current regulatory incentives to focus on errors of omission, but it is necessary that leaders influence their regulatory compact by painting an accurate picture of the new reality they are faced with. As power companies undergo changes, regulations must adjust accordingly.

5. **Investors' expectations.** In several countries, power companies and utilities are considered defensive investments with a relatively low risk-return profile. A large part of the investor base consists of long-term bondholders, insurance companies, pension funds, and other asset investors looking for a safe harbor. These investors provide little headroom for drastic increases in the risk exposure that is usually associated with new ventures, agile operations, and change with uncertain outcomes. Thus, if regulated utilities and power industrials have to underwrite riskier projects, the investor base likely has to change: newer and riskier projects require higher returns, which in turn have to seek a higher allowed rate of

return from ratepayers.[3] Innovation at scale will require changes to the current risk-return profile. Power industry leaders will need to address new challenges in investor relations and balance a multiparty dynamic with changing risk perceptions, investor returns expectations, and regulatory controls.

6. **Attracting skills and expertise.** To adapt and operate future business models, leaders will likely find that they need a different talent pool. Because the forces of change are not just affecting the power industry, in most markets the power industry will have to compete with other industries for these new talents and skills; in fact, the high demand for people with data and analytics skills is already evident. Attracting (and retaining) human resources depends on numerous factors including compensation, culture, engagement, and motivation. To draw high-quality employees, leaders will need to create new market-attractive career models. Many legacy talents and skills will not be relevant in the future. While companies can invest in upskilling personnel with some success, the reality is that many employees' skills will become obsolete. Leaders have an obligation to develop a transition plan that balances the business needs with the emotional and economic impacts of these staff transitions. This shift is particularly challenging given that in many countries, power companies were viewed as an "employer for life," valuing tenure over fit. Company leaders, especially the chief human resources officer, will become critical players in these efforts, and the human capital agenda—historically treated widely as an afterthought—will become core to emerging challenges.

7. **Few successful role models.** In the past twenty years, we have witnessed tremendous economic growth, especially in the technology sector, which has impacted all industries. The composition of the Dow Jones 30 that captures the blue-chip companies traded at US stock exchanges, for instance, looks vastly different than it did twenty years ago. The power industry during this period has also seen major changes with smart grid deployment and new business models reflecting plummeting prices

for solar, wind, and battery technologies. However, few power companies have defined and moved forward on an innovation path to reinvent themselves. The vast majority still have portfolios and modernization plans that are incremental and do not reflect or lend themselves to the changes that may be necessary in our VUCA environment. Also, with few success stories as examples, many companies have adopted a wait-and-watch approach. The adverse failures faced by some companies, which arguably moved too early or too quickly, also deter change and favor caution. NRG's David Crane, Engie's Isabella Kocher, and the downsizing of GE Digital are recent examples where a leader's early adoption of decentralization and digitization technology did not result in financial outcomes. The challenge for leaders is that leading their companies is not enough; they must also act as steward and custodians and rally other companies and stakeholders to generate momentum, and inspire change in the industry.

8. **Revising performance management.** "What gets measured, gets managed" is a cliché in management. Tracking and monitoring key performance indicators (KPIs) and metrics is essential for management. Digital tools and systems have created tremendous opportunities for improving the performance and efficiency of people, processes, and overall company operations. This success no doubt has created a proliferation of financial and operational metrics and KPIs, often to overwhelming proportions. It is not uncommon for C-level executive to receive a dossier of hundreds of KPIs every week. With new agile and flexible ways of operations, performance management is presented with its own challenges: namely, no available historical data basis or benchmark for comparison. Additionally, innovation and exploration activities require leading and forward-looking indicators. Most KPIs are usually backward-looking and largely based on accounting data. Hence, leaders and their boards are not as accustomed to using forward-looking measures in managing their enterprise and taking future-oriented decisions. New areas of focus such as customer experience call for new ways of measurement that are not easily observable, and relying

on survey-based scores such as the J.D. Power score may prove to be too noisy and ambiguous to guide management actions. A few utilities are looking into best practices from retailers and financial institutions regarding new ways of performance monitoring and tracking.

SUMMARY

The forces of global change call for significant rethinking and restructuring of the power industry if it is to remain viable. Industry leaders face monumental challenges as they strive to steer their companies through today's conditions—described as volatile, uncertain, complex, and ambiguous—into agile, nimble, and flexible organizations that can respond to this changing environment and succeed in the future. After noting that identifying and framing hurdles is a critical first step in navigating the new landscape, this chapter urges leaders to apply the model of the startup and to embrace seeding and scaling initiatives necessary to identify opportunities for survival and growth. Transforming a long-standing and matured industry to move and operate like a startup requires that leaders address a series of challenges, the hardest being embedding the startup-like operations within the core legacy organization, while simultaneously taking steps to preserve and protect the matured side of the organization, which is the source of economic value today. Leaders must balance these two important but seemingly contradictory priorities: maintaining business continuity alongside developing the ability to change and pivot quickly to meet the demands of the future. In the next chapter, we discuss the portfolio of solutions that leaders can bring to address these challenges.

NOTES

1 Research Briefings 1986: Report of the Research Briefing Panel on Decision Making and Problem Solving by the National Academy of Sciences.

2 A play on the phrase "Culture eats strategy for breakfast," attributed to Peter Drucker.

3 Ratepayer is a term used in North America for a regulated utility customer. In this book, unless otherwise mentioned ratepayers and customers are used interchangeably.

CHAPTER THREE

Components of the Solution

What Are the Ways and Means to Counter the Impacts of Change?

DOI: 10.4324/9781003353997-5

It is well known that most major organizational change efforts fail to deliver desired results, and we have no misconceptions regarding the odds of success in turning around and re-creating the power industry. Yet we have also seen that when the threat is existential—as it is to the electrical power industry today—organizations throughout history have had to transform. Recent lessons from the US military are a good example. When General Stanley McChrystal realized that the US was losing the war in Iraq in 2004, he understood that the Joint Special Operations Command (JSOC) had to change. To begin with, he had to question and indeed, overcome the prevalent belief that the JSOC was already good at what it did. He began by identifying and exposing current gaps and failures in coordination and synergy across various teams that were working in an insular and siloed way. He had to show that the teams were not equipped to outsmart the complex network of the adversary with the speed that was needed. General McChrystal initiated a change effort to move from the legacy hierarchical command-and-control model with centralized, authoritative, and top-down decision to a network of teams—what he called the "team of teams." As McChrystal wrote in *Team of Teams: New Rules of Engagement for a Complex World* (2015), "The Task Force hadn't chosen to change; we were driven by necessity":

> Little of our transformation was planned. Few of the plans that we did develop unfolded as envisioned. Instead, we evolved in rapid iterations, *changing—assessing—changing* again. … Over time we realized that we were not in search of the perfect solution—none existed.

This example illustrates where a leader's engagement and quest for a solution begin. Leaders first must understand the unique challenges facing their organization and then find approaches that are effective in addressing these challenges. Adapting to the changing environment is not optional; it is what distinguishes a successful company transformation from a failure. In this chapter, we highlight components critical to effectively countering challenges, and we provide a case study on planning in an uncertain environment.

CREATING AND PURSUING A VISION

Good leaders are devoted to fulfilling their missions. The challenge in a changing environment is to identify, articulate, and stick to these missions. The most common shortcomings leaders exhibit are lacking a defined point of departure for a course of action or, similarly, waiting for the fog of uncertainty to disappear before taking action. Both have paralyzing consequences. To avoid these pitfalls, a vision of the future is necessary. Most business leaders understand the need for a sharp vision to set the direction for their strategic plans and actions. Yet a vast majority finds the visioning exercise to be little more than a management ritual with limited effectiveness in serving as a robust guideline for personnel across the organization.

The most common issues with vision statements are that they don't paint the future picture in vivid details, they are not designed to address divisions across and down the organizational chart, they are shared only with senior leaders of the organization, leaving lower-level managers and staff without a clue, and they are too far removed from how the leadership is thinking about the future. Many vision statements are mere aspirations colored with metrics. For instance, deciding to become a more customer-centric company or pledging to achieve net zero does not paint a full picture of what the future under these vision commitments means. A vision is a subjective narrative of the created future that a company will live into; it provides a vivid description of how the company will conduct itself. Leaders can test and confirm their vision descriptions if their vision is clear and can be easily communicated to the teams to create a shared understanding. The following are a few questions to address when creating a company vision:

- Is it coherent—meaning do the various aspects hang together and make cogent sense?
- Is it fundamentally grounded in the company's values and is it consistent with the company's overall mission?
- Is it aspirational and credible—meaning that it is not easily achievable but realistic and not a pipe dream?

- Is the vision shared among key stakeholders in the company—meaning that leaders and decision-makers in critical positions are aligned and committed?
- Is the vision resting on a vivid description of the future—meaning that the narrative on the future vision paints a picture that is clear to all?

Leaders need to determine the process that works best to create their organization's vision. In our experience, companies that create detailed narratives that are carefully crafted into strategic plans and actions are most effective. The more specific the view of the future is, the more effective it becomes to rally the organization around it. A detailed vision also saves gross inefficiencies by weaning out those who do not align or are unwilling to commit to it. Further, it removes possible ambiguity and vagueness for the stakeholders, it keeps the organization aligned to a shared view of the future, and it helps in setting near-term targets and the course of action, which will be specific to the company.

STRATEGIC PLANNING FOR UNCERTAINTY

The capital-intensive and mission-critical nature of the electric power industry has engendered a tradition of a strong planning discipline that has been perfecting the prediction of future operating conditions. System planners rely on industry-leading load forecasting, and planning large building projects is a core capability for power companies. But the traditional planning skills do not serve well under conditions of uncertainty. Under uncertainty, plans need to be developed with incomplete or no historical information, and plans must allow for changes to respond to changing realities. Such planning requires shifting from an approach of predicting the future to preparing for it: preparing and garnering abilities, negotiating events that unfold, and capitalizing on opportunities. Such planning is particularly difficult for the power industry because it has to plan for the extremes. On the one hand, a dynamic environment requires planning for agility; on the other hand, mission-critical functions like real-time grid operations and control need to be planned in detail so that operators adhere to strict procedures to avoid chances of error. Identifying the functions and projects that need shorter planning cycles, agility, and flexibility versus those that need to be precise is an important leadership decision. Most industry leaders have grown up in Taylor's[1] scientific management culture, which separates planning from execution and works extremely well in a predictable environment but not under uncertainty. No amount of estimation and analyses will provide an accurate and precise forecast for the future under VUCA conditions.

To plan for agility, the gap between planning and execution must be bridged with the ability to make tactical decisions close to where the action is. Therefore, for these activities, plans must be constructed in a way that avoid the need for higher-ups to provide directions and approval. To achieve this level of planning—objectives, resources, and opportunities must be assessed and addressed differently from how they would be using a traditional approach. Specifically, a leader must focus on the following:

- **Objectives.** Even if specific details are uncertain or unknown, the leader must ensure objectives for teams and cross-functional groups at every level of the organization and codify what teams and individuals must do.
- **Scenarios.** The leader should consider multiple possible future scenarios in planning the course of action. If the end point for the future is not fully known, or is uncertain, the leader should set the plan in the right direction, rather than second-guessing or striving for precision.
- **Stakeholders.** The leader should identify internal and external stakeholders (including customers, regulators, and investors) and work out what, where, and how the plan will serve their interests.
- **Alignment.** The leader should ensure that linkages between vision, strategic options, and objectives are aligned at all times.
- **Measures.** The leader should develop data-driven evaluation criteria grounded in facts and logic that can be applied transparently and

consistently to make trade-offs (e.g., short-term financial results versus longer-term investments).

- **Review and update.** The leader should create criteria and conditions for review and what will require the revision of the strategic plans.

If companies do not have a planning process to account for uncertainty, leaders may need to broaden their strategic planning framework. Figure 3.1 compares two planning frameworks—the traditional framework (on the left) and the framework for planning in a world characterized by VUCA (on the right). Under VUCA conditions, different projects will require different planning approaches depending on where they map on the four quadrants. Our VUCA planning framework captures the notion of limitations, ambiguity, and uncertainty.[2] Planning under uncertainty can elicit paralysis—a tendency to do nothing, often driven by risk aversion and fear of failure. But doing nothing is usually not the optimal option. To handle uncertainty, planners engage tools like scenario planning[3] to understand complex future states. Scenario planning uses a set of alternative futures to analyze whether and how a company will be able to handle such situations. Scenario planning is often used improperly to guess which future states will actually happen, but that is not the point. The power of scenario planning is to optimize for robustness and resilience and to understand the extensive implications of possible future

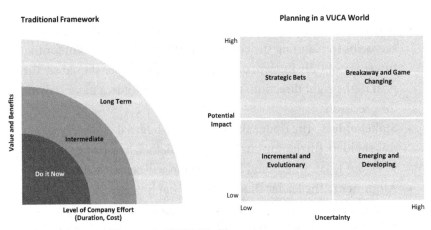

Figure 3.1 Shifting to Planning in VUCA World

states to the organization. Scenario plans also expose points of fragility and vulnerability and show how to prepare. Planning requires action, and the question is, *what is the appropriate scope and speed of action?* For such a planning framework to be effective, it has to be hosted by a company that is operationally designed for speed and flexibility. The processes need to be in place to enable the organization to flex its speed as necessary.

DESIGNING THE ORGANIZATION TO EXECUTE UNCERTAIN PLANS

OPERATIONAL TEMPO

To improve speed, a traditional power company must go through some specific changes—what in military parlance is called building "operational tempo"[4]—for the swift implementation of its strategic plan. Operational tempo in the organization is the number of events that the organization can respond to within a period of time. Speed is essential for operational tempo, but it must come from specific action. Observing, orienting, and selecting the events to respond to are critical components of developing operational tempo.[5] Not every area in the organization needs to develop the ability to move with tempo. Asset maintenance, grid control, and service restoration do not need high operational tempo but need to be efficient and rely on tested, proven methods that have been perfected from practice and repetition. In these areas, hundred-year-old Taylor's principles of scientific management and operational productivity are still valid.

Where the organization needs to be agile and likely to respond to the changing environment, where new initiatives must be developed and launched, operational tempo must be engineered. Such activities today include adopting new technologies, serving changing customer needs and wants, using data analytics for decision-making and operational optimization, and acquiring and training talent with specific skill sets. To make their organizations agile, leaders need to create a decision-making process that is based on an established strategic posture.[6] Once leaders assume a strategic posture for their organization, one that is consistent with its North Star and its organizational ability—and financial capacity, then the posture guides the decision-making that provides the speed. The task of developing the strategic posture is an intellectual exercise that lays the foundation for decisions that need to be made with agility and flexibility. Without this grounding, programs will be merely responding to unfolding events in an ad-hoc manner. The responses may give the illusion of speed, but they will most likely be disconnected from reality and are likely to fail, including in situations when no action is taken, and opportunities are lost. Taking no action can be a legitimate strategic choice, but only when informed and guided by a well-considered strategic posture.

A strategic posture is a strategic intent that includes approaches ranging from responding aggressively, preparing, or doing nothing. Strategic posture provides a common ground and drives the nature of the response that is executed at various levels of the organization. It keeps the organization and groups in alignment to respond when responding to changes without layers of evaluation, scrutiny, and approvals—thus bringing planning and execution together, often performed by the same entity. This posture is at the heart of what gives organizations their operational tempo.

We group strategic postures into two broad categories: the preparation postures and the execution postures. Preparation postures include (1) observe critical points, (2) analyze and assess situations, and (3) build capabilities. Execution postures are (1) shape or make the market, (2) experiment, and (3) undertake rapid deployment. As the terms suggest, preparation postures involve decisions related to observing and analyzing situations much like someone on the watch—observing and sensing the environment, and analyzing and looking for critical market or technological trigger points that will warrant some thoughtful action. Building capabilities such as innovation, asset management, inventory management, and customer services are longer-term actions but are part of the preparation to operate in the environment at a higher level of tempo and performance. Execution postures are decisions to act that include making limited bets in experimental technology projects and investing in shaping policies, market, and customer preferences. Both preparation and execution postures are active decisions that are integral part of the planning process.

Companies will need to adopt different strategic postures depending on the robustness needed for their environment and their ability to execute. Observing and analyzing may be relevant for new developments that are at the edge of the industry and have not yet become defining trends. Experimentation, often called *strategic learning*, is appropriate for potential game-changing technologies where small investments to test and learn provide a head start before making a large bet. Pilot experiments currently include green hydrogen, decentralized control, carbon capture usage and sequestration, advanced power electronics, and long-duration storage. Separate from the technical evaluation, part of the planning is to determine how quickly the organization can adopt and operationalize the technology at scale once a breakthrough threshold target is achieved. Conversely, the target indicates when the company must decide to pull the plug and reallocate its resources. Market shaping is another strategic posture that is not common in many industries, but it can be crucial for power companies. Influencing policies and regulations can create a market for game-changing technologies. Power companies have created intellectual property such as wind analytics and asset management, but few new concepts have been successfully monetized with a concerted market development. As new frontiers in customer products and services and data analytics develop, creating new market opportunities will become important in unlocking new revenue streams.

ORGANIZATION STRUCTURE

A strategic plan needs an organization structure to host it and give it life—otherwise, the plan remains an idea. Organizational innovation hardly gets the focus and attention that technological innovation does; however, in recent years, certain digital companies (e.g., Spotify) have innovated with novel organizational models that are proving to be successful in developing built-in agility and flexibility.[7] Startups that are much admired for agility in their early days often become centrally controlled organizations as they scale, and they lose much of that agility. If business organizations seek to innovate for speed, then they can learn from innovations on transitioning from command-and-control models to agile models in dealing with volatile and uncertain market environments with unfavorable odds.

The main organizational design challenge for agility is to balance autonomy with alignment.[8] Conventional wisdom says that these two attributes are a trade-off and that a leader has to sacrifice one for the other. An organization can respond quickly when it has autonomy and decisions can be made by those in action (the front line personnel) rather than by needing to ask for permissions and approvals from higher-ups. However, allowing autonomy is no guarantee of meeting the overall strategic intent of the bigger organization; a flat organization and autonomous group can lack coordination and lose sight of their shared mission. Hence, the actions at the front line must align with the overall strategic intent. Most organizations view alignment and autonomy as mutually exclusive, but alignment done in a specific way allows greater autonomy. Mission command in the military and companies like Spotify have designed organizations on this principle. Following are some basic principles on setting directions and tracking progress that can ensure alignment and autonomy.

SETTING DIRECTION, ALIGNMENT, AND AUTONOMY

1. Leaders (Level L) give their *command* (the directive to the leaders of the rank below [L-1]), which consists of their *intent* or purpose (the *what* and *why*, but not the *how*).
2. Leaders one level below (L-1) take the intent and work with their teams to draw up detailed tasks (*how* they plan to achieve the intent). Following the pattern, this cascade of intent and tasks flow all the way down the chain of command.

3. Focus is on the clarity of communication of intent. Carefully crafted statements are used. Uncertainty in the environment along with gaps in knowledge and understanding is acknowledged. Those gaps are used to guide the course of action.

4. Leaders' intent, directive, and command need to be as clear as possible, including understanding the order or priorities (for example, higher revenues, earnings, and reliability may be all part of the intent, but which is more important than the other).

5. Lower-level personnel are required to have a back-and-forth dialogue, and communication and resolution have to happen between the leaders and their reports so that the intent is clear.

6. Another component of the command, or directive, from the leader, is clarification of constraints. When execution plans are drawn, boundaries on resources, time, and budgets are clearly laid out. Teams are allowed to make decisions within the boundaries as long as their actions match the intent.

7. Alignment is achieved when the intent matches the execution plan through back-and-forth communication between leaders and teams. Autonomy is achieved when the teams decide on tasks in real time within the constraints so long as the task is aligned with the purpose. Thus, execution teams can deviate from execution tasks if necessary. They don't need to comply with the original plan, nor do they need to wait for another review cycle or get permission.

TRACKING PROGRESS AND RESULTS

8. Metrics and measures are needed to ensure that the intent and outcomes are achieved. Metrics and measures must be contextualized. For instance, the process for restoring electricity can vary considerably under different weather conditions. Capturing all the variables and scenarios is often impractical, and managing metrics alone without context can lead to harmful decisions or illogical behavior. These outcomes can be avoided by focusing on the overall intent and by inculcating trust in the team, judgment,

and other qualitative attributes that increase speed without the cost and bother of onerous reporting.

9. When progress reports are reviewed, the focus is on whether the intent was fulfilled rather than whether the original plan was followed. The goal is not compliance with the plan or even on process but on the commitment to action to fulfill the overall intent.

10. Measures, metrics, and KPIs should not be used as an end to themselves; rather, they should be used as indicators of where to look to fill gaps to elevate performance. They are articles of inquiry and not mere targets to meet.

11. The gaps, unknowns, and conditions behind the course of action need constant monitoring and periodic review to determine whether the current course or the immediate goal should be revised. Depending on the nature of change, alignment among concerned parties needs to be re-established.

BUILDING MOTIVATION, MORALE, AND CULTURE

12. Program leaders need to build teams equipped to perform and deliver desired outcomes. Teams need support to perform their mission/plan and to focus on action. A team's concern about hitting target metrics during action can reduce focus on the task itself and negatively impact performance.

13. Rewarding individual performance versus team performance must be minimized, as it can undermine the team's morale and performance.

Designing an organization on these thirteen points is not an overnight effort. The long-standing culture of compliance in the power industry that stems from the command-and-control model—which, as noted, requires personnel to follow processes, procedures, standards, and protocols—leaves little room for independent thinking to respond quickly to a changing environment. The predominant practice has been for people in power companies to do what they are told to do. A shift to an agile organization based on mission command ideas will be a process. Also, as stated earlier, there will be areas where the hierarchical command-and-control model will remain appropriate. For real-time operations of the grid, running a switching procedure in substations, or doing maintenance in high voltage lines, organizations need to follow tested, proven procedures practiced and perfected—that leave no room for mistakes. In addition, the new approach must be adapted to reality. If teams are inexperienced and risks are too high, then the approach should be scaled back to a more traditional command-and-control model, and as trust grows, leaders can delegate more. A balanced approach allows organizations to orient for increasing their operational tempo under a changing environment.

FORMALIZING A ROBUST PLANNING PROCESS

To meet organizational requirements, a formalized process for planning under uncertainty is necessary. Given the inherent nature of unknowns, data will not provide all the necessary insights for planning. Leaders need to plan based on a set of beliefs. Beliefs are not obvious truths but are hypotheses that are testable. They are based on evidence and compelling arguments that shape a shared view of the future. While beliefs should be enduring (that is, they should not change often), if the underlying assumptions change, then beliefs must be changed. Using beliefs to drive goals and plans prevents two issues: (1) it avoids the search for data and facts that slows down decisions and (2) it creates a shared sense of purpose for the organization. Figure 3.2 is an illustration of the process similar to the one used by Spotify. Two distinct sets of beliefs (hypotheses) are formed—one from the external macro environment (e.g., geopolitics, economy, and technology) and the other—from the data and insights collected internally by the working-level teams (e.g., demand, customer preferences, and weather). Using the two strands of beliefs, a set of initiatives or projects form the organization's project investment portfolio, which aligns with the strategic posture and balances autonomy

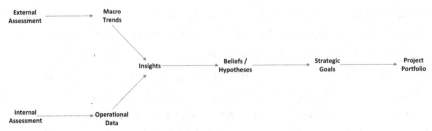

Figure 3.2 *Process to Provide a Sense of the Proportion of Unknowns and Risks*

with alignment necessary for the organizational team. These established, shared beliefs allow the planning process team to handle information gaps that are endemic in an VUCA world. Beliefs—along with the underlying information gaps and conditions (what needs to be true)—must be explicit and preferably have been discussed and written down.

CASE STUDY: PLANNING UNDER UNCERTAINTY

PART 1: BILL'S CHALLENGE

Bill has been recently appointed VP of grid modernization, a new position created to flesh out and execute the charter of modernizing his company's grid. Bill, with his previous leadership positions in regulatory, finance, and operations, is ideally suited for this position. He has been part of the executive leadership team (consisting of the CEO, CFO, and CIO) tasked with developing the vision for his utility to become a leader in decarbonization and to achieve 100 percent clean energy utility by 2050. Initial estimates worked out with the help of leading consultants indicate that the modernization of the grid alone—upgrading old assets (poles and wires), ensuring grid automation, and installing new sensing technologies—will require the company to invest $20 billion over the next ten years. The new position is a huge career boost for Bill, and he is excited to be spearheading the efforts. As a father, he finds profound meaning in his work as he believes he is making a difference for future generations. At the same time, Bill understands his challenge: translating vision into reality needs proper planning. He begins by working on a realistic plan.

The plans he has to develop are far different from anything he has done in his twenty-five years at the utility company. While the end goal is clear to him, he does not know how to get there. "There are so many unknowns. The technology landscape keeps changing. What if I invest in a technology and a better one comes along? What if policy changes?" he thinks. He wonders, too, how no one saw the solar PVDER market and how no one predicted that electric vehicles would grow so rapidly. And how could anyone predict extreme events such as the 2020 pandemic, which stalled the global economy and reprioritized the company's plan? Bill wonders how he can possibly draw a realistic plan. How will he know it is realistic? How will he get the organization to believe in, commit to, and execute the plan? He realizes that his assignment is not just another planning exercise. He knows of countless examples such as Kodak and Olivetti typewriters that could not reinvent themselves and survive when new technology reached a tipping point, despite valiant late attempts. So how shall Bill think about planning?

PART 2: A FEW MONTHS LATER ...

After several iterations of Bill's ten-year modernization plan with the executive leadership team, the board approves it. The plan is structured as a set of ambitions grouped under four flagship programs: (1) asset management to maximize operational efficiency, (2) investment in distributed generation, (3) field force effectiveness to drive productivity, and (4) digitalization of enterprise. The approach and set of actions of Bill took to develop his plan are as follows:

1. Identified the set of unknowns: Bill and his team agreed that over the ten-year horizon, the operating environment is likely to be very different if the past ten years are any indicator. There was no way to predict which technologies or policies will dominate the business.

He believed that cloud-based services will grow, cybersecurity concerns will increase, supply chains will be more integrated but also vulnerable, sustainability concerns will increase, and electrification of transportation and possibly of space heating will also grow. But precisely how fast and how much will be unknown?.

2. Created a detailed charters for programs that were familiar: In the area of asset management and field force productivity, Bill knows there is historical data to inform the future course of action in more accurate details. He can create a charter with intent and purpose and set some goals to achieve them. He sees a pathway to achieve productivity improvement of 20–30 percent using tested, proven, and predictable methods.

3. Set directional goals and not end goals for new programs: In two emerging areas, investment in distributed generation and digitalization, Bill has no reliable historical data or experience. He decides to assemble the team and set some direction instead of an end goal. He discusses with his peers and experts to determine some measures and metrics—for instance, the percentage of distribution circuits to be made ready for DER adoption and the number of functions, processes, and data sets that are to be digitized. He realizes that he can drive clarity with a stated strategic posture. So, he articulates that the plan will be to choose "readiness" and "execution" as the strategic posture.

The team should know and prepare enough so that deployment can be mobilized depending on how the market unfolds. The plan will require several sessions of strategic analysis to guide the next set of actions, including which experiments to conduct, when to wait, and what trigger points to monitor, and so on. "I don't have a plan based on predicting the future. It would be useless," he says. But he has clarity on what to do next, how to mobilize his team, and how to chart a course that is directionally accurate. He has rigorously documented information gaps and how wide they are, and he has drawn multiple perspectives from his team to ensure alignment around a set of beliefs that are formed from the available data and insights. His plan has specific next steps and measures for success. His approach shows the built-in flexibility when teams are encouraged to modify steps as long as their direction is consistent with the aim. Because he engages the team and gives them autonomy in making tactical decisions, Bill's plan shows alignment to strategic intent, as well as Bill's prudence in not taking a huge, risky bet, with large-scale deployments. In this process, Bill also identified that, because of the historical separation of the regions and of the functional teams, little trust exists among functional and regional groups. People down the ranks were reluctant to commit to timelines and spend levels. Detailed estimates came bloated and padded with contingencies. One individual shared privately with Bill,

> We do this because, if our spend goes over, even if it is due to events outside our control, there is a huge reprimand, and our bonuses are taken away. It is as if we failed in our planning or execution.

Bill knows building trust and morale will be necessary, and importantly to sustain it. He needs to enable his team to take initiative and promote commitment over compliance, directing over ordering, confidence over control, and respect over fear. He also recognizes that certain people will not adjust or fit into the new model, and they will have to be moved out. Bill realizes that his first order of business is to create trust in the team so that it can make tactical decisions, solve problems, and apply itself to plan and execute. When he briefed the board and C-level executives, Bill's transparency and evidence-based logic were met with initial discomfort. "How can a plan leave so many unknowns?" a member queried. There were a series of back-and-forth sessions both in groups and individually. Bill made the effort to understand and iterate, and he was finally able to obtain alignment with the board, his senior team, and his core execution team on intent, constraints, and how the team would accomplish the aim. The mission was on.

LEADERSHIP ATTRIBUTES

L eading in a VUCA environment calls for a heightened presence of certain leadership attributes. These attributes may seem obvious to seasoned industry leaders, yet in practice they are often overlooked or slip into the background:

- **Dealing with uncertainty.** Coping with uncertainty becomes increasingly difficult using existing models and practices, and leaders must have the ability to make decisions with incomplete information, being aware of ambiguity yet able to navigate unforeseen events that may arise, all while exuding and inspiring confidence and clarity in their organization.

- **Grounding in reality.** Leaders need to be realistic. There are many reasons to deviate from the program plan and objectives and get trapped into wishful thinking. The VUCA environment often presents counterintuitive situations, and it is easy to default to taking decisions from the gut. More than half a century ago, the polymath Herbert Simon coined the phrase "bounded rationality," meaning that instead of acting perfectly rationally, leaders are humans who are driven by limits of cognition.[9] "When emotion is strong, the focus of attention may be narrowed to a very specific, and perhaps transient, goal, and we may ignore important matters that we would otherwise take into account before acting."[10] To limit the effect of bias, leaders need to place a high value on data and facts grounded on realistic expectations, as they design and craft their plans, budgets, and performance.

- **Deciding based on evidence.** Leaders know that decisions need to be based on evidence. In a noisy world affected by social media, opinions often masquerade as facts and can lead to wrong decisions. Consequently, leaders need to be intentional in their fact gathering and they must be on the alert for such traps; they must not be fooled into beliefs and judgments that can be harmful to their program and people, themselves, and their company, particularly regarding contentious topics, such as climate change, sustainability, and retiring fossil fuel plants.

- **Cultivating trust.** Often an overused word in business, *trust* is critical, and leaders need to be intentional about building it, and developing a mechanism to identify and restore trust gaps, when they emerge. Trust must be nurtured and maintained if leaders are to drive the program toward successful implementation. Trust is only cultivated when leadership acts consistently to promote behaviors that build trust and censures behaviors that erode trust in organizations. Program leaders must be ruthlessly committed to deal with personalities that perpetrate a low-trust environment and must strive to build high-performing teams based on trust. Trust is reinforced when teams can share responsibility, can delegate without fear, and can work consistently as an integrated whole.

- **Building an ownership culture.** An ownership-based culture is not just about having equity ownership in the form of stock options. Ownership attributes include having freedom and autonomy in making decisions and taking personal responsibility for outcomes. Ownership shows up in the drive for the shared goal and is realized when there are shared values, pride in the organization, and belief that everyone's contribution and participation matter.

- **Relying on scenarios and not on predictions.** In a VUCA world, the possibilities of the future cannot easily be forecast from the present; in fact, many signals are easy to ignore when predictions are made using probabilities. Leaders need to accept that the best predictions will likely not match the unfolding reality—despite the skills of managers, predictors, and estimators. A better approach is imagining and conjuring a specific scenario, which shifts the focus to finding ways to navigate through that situation, and mitigate possible events should it arise. Working with scenarios requires skill, but tools and methodologies have been developed to enhance organizations' ability to imagine different company scenarios and responses.[11]

- **Allowing divergent thinking.** In an uncertain world, the future can only be imagined. Leaders will need divergent thinking to see beyond the industry boundaries to determine what might be relevant to their decisions today. Divergent thinking encourages open-ended questions, brainstorming, and ideas. It is a complex process because at some point divergent ideas need to converge to a set of manageable items to allocate resources and to create objectives and measures to track progress.[12]

- **Overcoming groupthink.** Leaders need a mechanism to support a culture where teams don't succumb to following an idea without much examination. Often this misstep happens in benign ways, such as when the most vocal person in the meeting sets the agenda or when the team blindly follows the "star performer" or authority without questioning. Sample mechanisms to avoid such traps include the style of communication introduced by Jeff Bezos in Amazon and by Steve Jobs in development sessions at Apple. These specific practices may not be easily transferable to other organizations, but similar structures include assigning roles, designating a devil's advocate, and creating contrarian teams like the red team/blue team adversarial combination used in the US military. Groupthink can also make certain topics taboo. In one company, we observed that the acceleration to solar and wind resources was so strong that no one was willing to examine economic feasibility and other constraints. A few years later, the company faced capacity shortages because the obvious pitfalls were never brought up and examined.

SUMMARY

To navigate and chart their organization's course in the current environment, leaders need a strategic planning process. The strategy cannot rely on historical plays of incremental productivity and efficiency improvements, on synergies from consolidation and M&A, or even on targeted automation. Today's navigation requires a fundamental shift both in understanding future challenges and in creating a strategy. Because of increased uncertainty in the future, a strategy must encapsulate flexibility and robustness and must realize the company's vision in the form of a common set of action choices. This chapter describes how to create a strategic posture despite limited information and knowledge. A strategic posture can range from active preparation to real commitments, and it should allow for a flexibility of choices that can be taken with the appropriate velocity and effectiveness. The strategic planning process must be supported by an organization that is designed for operational tempo to provide agility and flexibility. This process will involve innovation in how leaders communicate intent and direction that cascades through the organization and balances autonomy with alignment. To perfect this process, leaders must draw on specific leadership attributes based on trust, beliefs, and reality. In addition, two areas will be key to navigating power companies through this time of change: innovation and customer engagement. These areas are discussed in the next two chapters.

NOTES

1 In 1909, Frederick Taylor published "The Principles of Scientific Management" which formed the foundation for many 20th century management practices in the industry.

2 Readers are encouraged to see Martin Reeves, Knut Haanæs, and Janmejaya Sinha, *Your Strategy Needs a Strategy: How to Choose and Execute the Right Approach* (Boston: Harvard Business Review Press, 2015).

3 See Pierre Wack, "Scenarios: Shooting the Rapids," *Harvard Business Review*, November 1985, https://hbr.org/1985/11/scenarios-shooting-the-rapids.

4 George Stalk and Sam Stewart, "Tempo and the Art of Disruption," *Boston Consulting Group*, February 28, 2019, https://www.bcg.com/publications/2019/tempo-art-of-disruption.

5 Stalk and Stewart, "Tempo and the Art of Disruption."

6 Readers may find similarities between operational tempo with the military's "OODA loop" concept, meaning the process of observing, orienting, deciding, and acting. Strategic posture planning and its relation to strategic business analysis were formulated by Ansoff; see H. Igor Ansoff, *Implanting Strategic Management* (New York: Prentice Hall, 1984).

7 Michael Mankins and Eric Garton, "How Spotify Balances Employee Autonomy and Accountability," *Harvard Business Review*, February 2017,

https://hbr.org/2017/02/how-spotify-balances-employee-autonomy-and-accountability.

8 See Stephen Bungay, *The Art of Action: How Leaders Close the Gaps Between Plans, Actions and Results* (Boston: Nicholas Brealey, 2010), and Stephen Bungay, "Mission Command: An Organizational Model for Our Time," *Harvard Business Review*, November 2010.

9 Herbert A. Simon, *Models of Man, Social and Rational: Mathematical Essays on Rational Human Behavior in a Social Setting* (New York: John Wiley, 1957).

10 Herbert A. Simon, *Administrative Behavior: A Study of Decision-Making Processes in Administrative Organizations* (New York: Macmillan, 1947).

11 Kees van der Heijden, *Scenarios: The Art of Strategic Conversation*, 2nd Edition (John Wiley and Sons, 2005).

12 Mark A. Runco and Selcuk Acar, "Divergent Thinking as an Indicator of Creative Potential," *Creativity Research Journal* 24, no. 1 (2012).

Addressing Three Critical Historical Gaps

CHAPTER FOUR

Innovation
What Are the Areas of Innovation?

DOI: 10.4324/9781003353997-7

The power industry is not known for innovation, but innovation is vital if the industry is to renew and stay relevant. Customer demands, sustainability and climate concerns, and the role of data and analytics have created powerful incentives for industry players to focus on innovation—innovation not just as a handful of ad hoc pilots or proof of concepts, but in terms of completely new business models that alter how companies do business and create commercial value. Innovation on this level is inspiring as well as challenging, particularly in the power industry, which holds large investments in assets, and where the services are critical to people's lives, and hence are closely monitored and regulated.

Innovation is linked to change, and when an industry is set in its ways for many years, change must be pursued carefully and systematically because innovating in one area may have undesirable impacts on another. While there is much use of the term "disruption" in the industry, not all innovation activities need to be transformational and disruptive. Some innovation focuses strictly on incremental improvements such as productivity and efficiency gains required to sustain the business. Innovative ideas also don't need to be all original and sourced from designated R&D centers; innovative ideas can come from anywhere. People with experience who have a capacity to spot patterns and reflect can spawn ideas for improvement. Fresh ideas can also be adopted from other industries that look at problems differently, specifically from those industries who have experience in driving innovation processes in a long-standing business. This chapter addresses the steps leaders must take to bring innovation to their companies—steps such as prioritizing company structures, building an innovation portfolio, organizing for innovation fitness, overcoming inertia, securing funds, and investing in digital platforms—which eventually will lead to the creation of an enduring innovation culture.

STRUCTURE BEFORE CULTURE

One of the doyens of modern management, Peter Drucker's famous words "culture eats strategy for breakfast". have tempted many managers, executives, and consultants to place an overabundance of focus on organizational behaviors and culture. Similarly, "building a culture of innovation" is a theme that percolates and dominates discussions in boardrooms and conference spaces. It's therefore no wonder that leaders are surprised when we advise them to focus on the structural building blocks of innovation before addressing culture. The structural components are the processes through which innovation projects are selected and constructed into a portfolio, how organizations are designed to aid the necessary communication and coordination regarding change, and what incentives drive desired actions and avoid unwanted practices. These structures strike a balance between short-term results and longer-term outcomes, and between centralizing versus decentralizing innovation activities. They also govern capital allocations, affect the alignment of investors and regulators, and shape the mobilization of the organization through the innovation phases—from ideation, scaling, and adoption within the core or legacy business.

Our argument for addressing structure first is not to say culture does not matter, but the evidence is strikingly clear that it is structure that is more important for leaders' focus at the onset of implementing innovation. Long-standing cultures do not change easily, particularly when the legacy business continues to run—and, in fact, in most cases funds the innovation efforts. Without addressing tangible matters of structure, the dominant long-standing culture that the business needs for present survival will overpower any push to change. In contrast, companies that have focused on hard structural elements to drive innovation have been more successful and over time have instituted new behaviors, norms, and practices that eventually led to cultural change in their organization.

BUILDING AN INNOVATION PORTFOLIO

Most leaders understand the importance of managing innovation initiatives as a portfolio rather than as a set of ad hoc activities. The portfolio must align innovation initiatives with the company's strategic direction. One way to ensure alignment is to work backward from the company's envisioned future. If a company has pledged emissions targets by a certain year, then that goal should drive the innovation portfolio along with similar pledges and milestones. Quality innovation comes from good ideas. Good innovation ideas can come from anywhere. Along with the traditional sources of internal functional, developmental, and operational leads, ideas can come from practically any part of the organization, including suppliers, vendors, customers, and other external sources. Constructing a robust portfolio requires a screening process that channels ideas sourced from a variety of origins through a set of selection criteria, such as fit with strategy, fit with innovation goals, feasibility, and funding needs. A balanced portfolio allows for risk diversification and maximizes the chances of success of the overall portfolio.

In legacy companies, it is important that the innovation portfolio targets specific aspects of the business because not all innovation projects are the same. Projects can be broadly grouped into those that target incremental improvement of the legacy core business and those that call for a fundamental change in the business model. Employing new and improved electric insulators with advanced materials, replacing transformers with new materials that lower power losses, or automating a mix of manual or spreadsheet-based processes with off-the-shelf software are innovation initiatives that improve the core business. They require vetting, testing, and validation, and they are important for boosting and sustaining current performance. In contrast, developing advanced power electronics technology that has the potential for deploying DC transmission, inverter switching, and/or microgrids is an example of fundamental change. Such innovation ventures are speculative and have transformative potential. They are uncertain, and risks can be enormous when viewed through the lens of current business. Separation of innovations that result in incremental change from those that are transformational in the innovation portfolio allows for the two perspectives to coexist and yet managed differently. A potential game-changing visionary innovation may also be a drag on a company's free cash flow in the short term. But by separating them, both the legacy business and the breakthrough innovation project(s) can be managed using different criteria for capital allocation, performance measures, timing, and value expectations.

SELECTING THE PORTFOLIO

A successful innovation portfolio offers a critical mass of projects that deliver on the company's strategic goals. Not all projects in the portfolio will succeed; what matters is that the net returns for the successful projects more than compensate for the failed projects. This approach is similar to how venture capitalists invest in early-stage ventures. But in practice, when funding has to be provided from a matured business by taking a hit on current profits, the venture capitalist approach is rendered irrelevant and groups devolve into using familiar risk thresholds and hurdle rates, filling their portfolios with projects that have a sure chance of success. Most legacy power companies with a sustaining core business are naturally biased to improvements in the core business operations, yet the fact remains that there is a limit beyond which incremental improvements do not make a major difference in the bottom line. That's the time for a fundamental change. Many companies go through rounds of piecemeal automation and cost takeout exercises until the organizations are left with no slack, lose resilience, and accumulate a huge innovation debt to make a step jump in performance. The old pillars that supported the legacy business will not be able to support the new structures needed.

Leaders can play a critical role at this point by orienting their organization toward the future and verifying that the innovation portfolio and actions taken today are fully in line with the company's aspirations. When portfolios appear misaligned with strategy it is either because of a lack of institutional support to allow projects of certain risk levels or because of an absence of genuine will to meet strategic ambitions. Additionally, many portfolios are filled with projects that are unrealistic for the organization.

The display in the theater of innovation attracts media and publicity, and even the prying eyes of business academics, until everyone realizes the impracticality of such actions and pledges. Once these commitments start to unravel it is only a matter of time before targets are missed, a crisis brews, and trust and credibility are eroded. Leaders must seek to align the company's practical boundaries and constraints such as timelines and budgets with expectations from innovation with its North Star.

When innovation is about a fundamental shift in the business with game-changing outcomes, business leader accountability is even greater. Where possible, leaders must set the tone and push their teams to think beyond average portfolio mix benchmarks, beyond the ratio of sustaining versus breakthrough initiatives, and to design portfolios that support the strategy. Given that the current wave of changes in the industry demands a higher proportion of such projects compared with past demands, leaders must be directly involved in leading the innovation agenda.

Once innovation portfolios are in place, they must be actively managed, which means individual initiatives or projects must have plans and controls with metrics and KPIs to track progress. When a project runs off track, appropriate interventions are necessary, and if projects are failing with no real learning for the future, they must be abandoned, and resources must be reallocated. Many programs take too long to cut the losses. To determine progress, KPIs and metrics do not need to be all quantitative; qualitative metrics are also useful and, in many cases, more meaningful in determining success. Customer experience outcomes, sustainability choices, technology improvements, and digital experiences that the company wants to provide can be more clearly stated in the qualitative form with supportive quantitative and financial results such as potential growth and margin enhancements. As many companies are making clean energy and emissions reduction pledges for the next two to three decades, it is on leaders to be accountable for aligning their innovation portfolios to these pledges and commitments.

When selecting projects for their company's portfolio, leaders should be aware of functional and departmental biases. What drives the composition of the portfolio today is mainly directed by current understandings of uncertainty and risks along with funding and resource requirements. Unlike venture capital funds, legacy companies generally have to allocate funds from their core business to fund innovation. The adoption and scaling of projects depends on expected business attractiveness and the operational changes required to bring the innovation into the core business. What makes an innovation initiative appealing today is when the view of the future market attractiveness is high, and when the degree and complexity of operational change is low. However, what might be financially attractive or easy from today's investment perspective may not be all that attractive when projected to the future. Factors beyond the control of the company pose additional uncertainty. Many innovations may require new materials, new computational breakthroughs, complex data collation, and mass market adoption for scale economics. Electric vehicles, rooftop solar PV cells, and artificial intelligence/machine learning (AI/ML)—the innovations creating transformation in the industry today—were considered nascent and risky bets only a few years ago. Similar impacts in the near future are now envisioned for long-duration storage, new sustainable technologies like carbon capture utilization and storage (CCUS), and clean fuels like green hydrogen. But uncertainties are high regarding economic and commercial scaling. Probing the innovation portfolio using current feasibility and future projections, as indicated in Figure 4.1, will lead to a richer and more robust portfolio.

Initiatives and projects within the portfolio must be scrutinized constructively. If such scrutiny is a new practice for the organization,

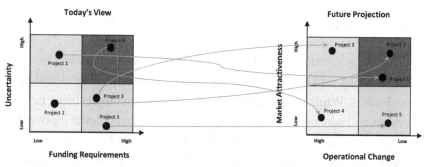

Figure 4.1 *Innovation Projects Current View vs. Future Projections*

leaders can oversee the introduction of norms and behaviors that allow for diverse views and fact-based assessment. When dealing with many unknown variables, teams often resort to asserting their points of view without providing adequate evidence—a process that can become destructive if team dynamics are reactive, passive, or even conforming (as when members fear appearing adversarial). In setting norms, leaders can question underlying assumptions, and they can introduce "creative tension" to reach a deeper understanding through debates and discussion among core stakeholders and relevant participants. Creative tension happens when there are several viable ideas and a choice has to be made, but choices are made after fact-based discussion. Jeff Bezos, the founder of Amazon and its former CEO, calls this process "disagree but commit." Netflix has its "debate, decide, and do." Innovation does not happen without disagreement; at the same time, endless debating is meaningless. Decisions must be made, and teams must get onboard with them. Finally, crafting the innovation portfolio is an ongoing process reflecting innovation decisions and future objectives; it is not a "one and done" activity, but a continuous activity of modification and adjustment as situations change. Depending on the nature of the project and its dependencies on external and internal factors, certain parts of the portfolio may have to be reviewed periodically. Such reviews must cover both the business case for the project itself and also the impacts on the overall portfolio—and must be aligned with the company's North Star.

MANAGING PRODUCTS, SERVICES, AND BUSINESS MODEL INNOVATION

Today's innovation agenda demands a wide mix of innovation initiatives, and some of these initiatives will require close attention from leaders. While product and services innovation targets bring new products and services at scale to the company, business model innovation involves a wholesale change in how the company does business—how it serves its customers and creates stakeholder value. New business models

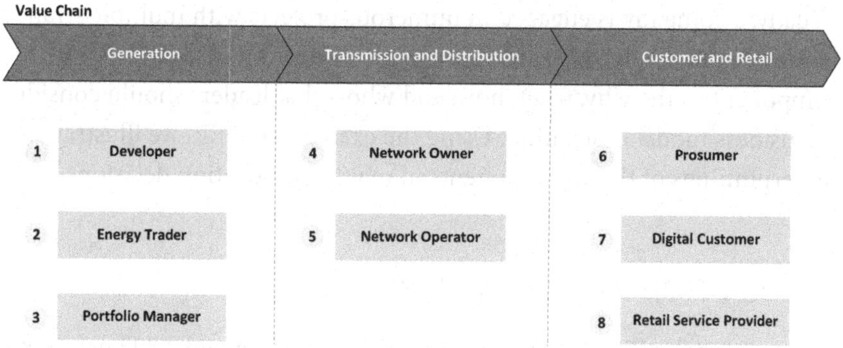

Value Chain

Generation	Transmission and Distribution	Customer and Retail

1	Developer	4	Network Owner	6	Prosumer
2	Energy Trader	5	Network Operator	7	Digital Customer
3	Portfolio Manager			8	Retail Service Provider

Figure 4.2 *Example of New Business Model Possibilities*

can emerge on the back of a new technology breakthrough or from major policy changes and changing customer needs and demands. All of these forces are at play in the power industry today. Figure 4.2 is one such representation. The emergence of digital platforms provides opportunities to serve new markets created by growing demands for sustainable and decentralized energy. A major business model innovation in the industry creates an existential threat to the legacy businesses that are too slow to change, and simultaneously, it creates windfall value for those that lead the change and secure dominant market position. The evidence is clear: companies that invested in renewables early on and innovated their operations from back-office systems to new project development are valued much higher than their legacy peers, due to higher scores in resilience and future proofing for growth.

It is possible to be a late entrant in the product and service space, and it is even desirable in certain cases to be a late mover to avoid incubation and gestation risks. But business model innovation is hard to emulate because it consists of changing an entire web of activities, skills, processes, systems, roles and responsibilities, and revenue models. Given the scope and scale of business model innovation, leaders must ensure that the innovation portfolio is tuned to the range of possible future business models. Projects must not just concern about individual products, services, or solutions, but must also take into consideration the relevance of the business and its future performance.

Usually, a company is engaged in numerous projects with multiple strands of innovation involved. In such a dynamic environment there are key components—the why, what, how, and who—that leaders should consider in business model innovation. Using the example of DER, we illustrate the underpinnings of these components in making innovation decisions.

VALUE PROPOSITION (WHY?)

If a traditional utility based on centralized generation and T&D networks decides to enter a decentralized DER business, then the company's new value proposition must articulate why DER will be attractive to its customers. Beyond the rationale for reliability, affordability, and safety, other reasons might include meeting policy goals for sustainability and providing customer choice to opt into clean energy resources. An estimate of the future value must be big enough to attract relevant players, suppliers, and partners. New operational characteristics of distributed generation and must inform and outline the path to rollout and scaling. The plan must also examine the factors critical to running a successful new business, along with the legacy business. Innovation efforts must explore the value of ecosystems, partnerships, and other new modes of engagement with markets and regulations that will get unlocked to the benefit of the company.

ACTIVITIES (WHAT?)

Activities refers to what the business will do. A DER business consists of physical assets from providers that supply solar panels, storage devices, inverters, switches, or "box-ed" systems like microgrids that perform their designed function. The network provider, often the incumbent utility, provides the network assets to connect the DER to the grid. If the DER is customer-owned and customer-located, customers are responsible for production. Site controllers conduct peak shaving, resource time shifting, EV charging, voltage/frequency support, load support, and backup. With aggregated DER, a DER management system (a grid control software platform), is used that also manages the distribution system. Other utility systems—outage management system and supervisory control and data acquisition systems—provide grid stability, frequency and voltage control, peak load management, and trading. Effective operations also require planning and modeling, weather forecasting, scheduling and optimization of DER asset operations, circuit and feeder load relief, and Volt/VAR management. To enable all of these activities, innovation—both in form of incremental changes to current operations and in form of new capabilities—is necessary.

STRUCTURE (HOW?)

DER activities are typically organized into functional and regional groups. Functional groups can be systemwide or process- or function-specific, such as behind-the-meter generation, DER aggregation, microgrid operation, power quality monitoring, protection coordination, risk management, and data management. The future business will depend on how these organizational groups (virtual power plants, aggregators, regulators, system integrators, and distribution system operators) are arranged to work in an integrated way, either through formalized commercial contacts, service agreements, or interface specifications.

GOVERNANCE (WHO?)

The future DER business requires new roles, responsibilities, and decision rights. Various options for the future may emerge, but finding and choosing an effective governance model will be key to the success of the business. In many jurisdictions today, the role of the distribution system operator is changing and requires regulatory authority over DER control and management. Policies and market rules are also changing, making innovation inevitable. There is no question that a successful future business model will also need a governance structure to manage the innovation activities, people, and systems to adapt to these changes.

ORGANIZING FOR INNOVATION FITNESS

Driving business model innovation of any kind is a complex process, and leaders must strive to install innovation teams that take a systemic view of all the components and their linkages. Whether it is choosing partners, architectural patterns, or customers—or even changing processes to improve speed or costs—trade-offs must be made. Also, because an idea is optimal for one group or function may still be suboptimal for the entire system; leaders need to ensure that ideas are appropriately vetted. A successful innovation portfolio needs an organization that is fit to deliver. Organization is a people matter, and it is critical that the organization design lends itself to innovation for the innovation to be successful. Internal communication is also key; messaging to people within the organization must be crafted with a level of specificity and rigor that is clear, relevant, and efficient. To achieve such messaging, the organization must have well-defined hierarchies, reporting lines, spans and controls, and roles and responsibilities. Leaders need to be aware that certain organizational designs facilitate actions more easily than others; certain designs can lead to a lack of fit between innovation strategy and structure. To test for organization fitness, leaders must answer the following questions:

- What are the organizational entities that are best suited to take ownership and execute specific innovation initiatives within the set of existing policies and KPIs?
- Are the right leaders and teams—meaning leaders and teams possessing the right skills, acumen, and commitment—in place to meet the initiative's goals?
- Does the current structure provide decision rights for the entities entrusted with innovation so that they can execute their mandate effectively?

Most companies discover gaps when answering these questions. As stated earlier, not all innovation initiatives are the same; some may have a closer fit with the current organization than others. People involved in innovation projects are interested in what they need to do and how their performance will be measured. If the tasks for the initiatives are not clear, and if accountability targets are also not clear, then a breakdown in organizational effectiveness will occur. We discuss three key organizational fitness areas—communication needs, measurements and goals, and incentives—in greater detail in the following sections.

USING COMMUNICATION NEEDS TO DESIGN ENTITIES

To close organizational gaps, the obvious question is whether a new entity needs to be formed or whether the legacy organization can be restructured and reorganized. Will new lines of communication and engagement be necessary to realize the innovation's outcomes? If the processes of ideation, development, testing, scaling, and adoption align closely with the existing means of operation, then the existing structure may serve well. However, if the current hierarchy creates inefficiencies and bottlenecks, then the legacy structure is not suitable. Furthermore, if the innovation scope is broad and benefits multiple functions, a centralized innovation program may be more appropriate than initiatives that benefit only one group. If success depends on candid, open exchange and rapid experimentation, a rigid formal and hierarchical structure with formal channels will not be appropriate. Drawing from Clayton Christensen's influential work on innovators' dilemmas[1] and innovators' solutions, legacy companies can create units separate from the parent/core business to address the demands of innovations. Examples of such new entities include National Grid Ventures, corporate startups like RWE New Ventures and Innogy New Ventures, energy accelerators like the Free Electrons accelerator program, and equity investment funds like Energy Impact Partners. Over time, three common structures have emerged:

- **Incubators.** Incubator entities are separate divisions—or separate units within divisions—whose sole job is to develop innovations and hand them to the operating divisions or units.

- **Accelerators.** Accelerators are dedicated sponsored groups that seek applications from entrepreneurs to speed up the business. Accelerators provide different facilities from funding to physical space, including providing laboratory and test facilities.[2]
- **Demonstration centers.** Demonstration centers are mock setups to show how innovation prototypes work or are expected to work. Demonstration centers were popular in smart-grid developments in 2008–2010, and many use the latest audio and visual technology, augmented reality, and virtual reality to test scenarios and simulate certain behaviors.

In companies with similar structures, these groups cover a broad range of topics such as electric vehicles, the Internet of Things technology, big data, shifting from fossil fuels to renewables, demand-side management, demand response, and advanced distribution management. The growing popularity of these models does not mean that creating new entities and keeping them at arm's length of the legacy business has to be the default for innovation. The novelty of the innovation effort and the popularity of using separate centers can easily steer leaders into creating innovation entities. However, the existence of such entities alone does not ensure organization fitness. An effective innovation organization requires that leaders develop a deep understanding of the nature of communication and information exchange needed to drive a successful innovation initiative; in fact, this understanding is the highest priority to inform appropriate decisions.

ARTICULATING HOW INNOVATION TEAMS WILL BE MEASURED

Innovation teams and even entire innovation entities are often given generic or vague measures as targets. Without specific measures, teams are not able to translate the innovation purpose into the activities teams perform on day-to-day basis. For example, if a team is charged with developing new ways to elevate customer experience, then leaders must align the team's expectations to the measures (KPIs) that need to be delivered. Specific KPIs such as reduction in cost to serve, increase in usage of lower-cost channels, increase in first-time problem resolution, increase in new users, are necessary. Similarly, for a brand-new sustainability initiative, such as one exploring the feasibility of CCUS or green hydrogen, it is not enough to simply state the intent in generic terms. Specific targets (such as percentage reduction of costs within three years) provide a tangible measure to align with, focus on, and track efforts against. If forward-looking measures are not easily determined, as is the case with many breakthrough programs, interim objectives must be crafted with benchmarks for measuring progress. Many companies employ innovation councils or new venture committees that meet regularly to address/review issues and coordinate. Measures give these bodies a reference on which to base their decisions. Depending on the cross-functional nature and strategic importance of an innovation, these committees consist of senior-level executives representing different functions, or even of executives from external organizations. Leaders may also set up specialized networks that focus on thought leadership, solution development, and exchange of know-how. The various vehicles of innovation have their own purpose and allow the innovation teams to work on their mission.

ALIGNING INCENTIVES TO MEASURES

People respond to incentives, and incentives often drive people's behavior in organizations. Incentives are not necessarily monetary; in many cases, nonmonetary incentives such as recognition among peers, opportunities for professional learning and growth, or participation in a high-impact venture are motivating. Different incentives drive individuals differently. In one situation, an executive of a leading power company decided to offer high performers of his innovation team a nonmonetary incentive because the executive did not have the budget for a bonus. Instead, he offered to have a one-on-one dinner with the employee and his or her family. While some employees liked the dinners and felt rewarded, the incentive fell flat for others who were incentivized by money alone.

ADDRESSING THREE CRITICAL HISTORICAL GAPS

In addition, unspoken incentives and disincentives may be at play. For example, many employees did not join the team because they were concerned that doing so would preclude their tenure in the core legacy business and hurt their promotion chances. The uncertainty of the initiative came across as a high personal career risk without much upside. Such situations are common, and leaders need to ensure that innovation teams functioning within the legacy organization or within a new entity are clear on incentives.

Spans and layers in the organization may not appear to have direct relation to incentives, but it makes a difference. Broader span and fewer layers reduce hierarchy and steer individuals' focus on the innovation problem rather than on networking and projection activities to seek that next promotion. People balance their efforts on their work against the "return on politics," which can incentivize certain behaviors. When leaders don't pay much attention to incentives or assume that the mission itself will self-select the right individuals, they run the risk of attracting individuals whose incentives are misaligned with behaviors and actions necessary to drive the initiatives. Designing incentives outside the norm is an arduous task and can often become politically charged. Hence, leaders must make it a priority to ensure that their choice of incentive is not just about paying lip service or communicating in management platitudes; the incentive must have real value, be meaningfully crafted, and be diligently executed.

OVERCOMING INERTIA AND LEADING THE TRANSITION

Long-standing companies that are set in their ways of operations are naturally biased toward the status quo. Organizational inertia resists and ultimately kills innovation. It damages creative ideas, deters exploring options, and diminishes the range of possibilities. Overcoming inertia requires going beyond finding insights and ideas; it requires strong leadership, and thoughtful strategies to break it. To do so, leaders must first get to the root of what drives inertia. We've identified four key causes of inertia:

- First, a **lack of familiarity** of what the future operations will be. Many new technologies are evolving, and the full range of impacts on work processes, organization groups, individual roles and responsibilities, and customers and stakeholders is not known.
- Second, people are **uncertain about their careers** and what a business model change means for their future, jobs, livelihood, and position in the company. Uncertainty bodes fear and anxiety, and basic needs for security and stability clash with the optimism and promise of new technologies.
- Third, the **time it takes to change company culture** gives the impression of sluggishness and the fact that things are not changing despite all efforts. Such perceived slow progress deters enthusiasm and challenges the motivation to stay persistent.
- Fourth, the **uncertainty of scaling early efforts** raises doubts and skepticism on whether the efforts of today will ever bear fruits and create the desired outcomes at scale.

The nature of regulations in the power industry can also contribute to organizational inertia. Table 4.1 shows a list of regulatory impacts on innovation. Regulations are often hard to change. Regulatory controls are in place to check for discretion and to avoid wasteful and reckless spending on risky innovation. But extreme reluctance to regulatory or policy change often dampens risk-taking. Often companies use regulations to promote "regulatory capture," an idea advanced by George Stigler in 1971, meaning that industries that are regulated are the ones that use regulation to their own benefit, including promoting status quo, and often counter to very intent of the regulation.[3] In 2013, during the early days of DER, Edison Electric Institute, a trade group of US investor-owned utilities, warned the industry of the "utility death spiral,"[4] which sounded an alarm throughout the industry. While many utilities in the US dismissed the idea, many others stood on the fence and influenced their regulators to preserve and protect the existing operation.

Table 4.1 *Examples of the Effects of Regulation on Innovation*

Feature of Regulation	Effect on Innovation
Entry restrictions for new companies	• Reduces competitive pressure on utility to innovate • Natural monopoly structure favors large-scale technologies
Regulatory lag	• Deters innovation as it takes longer for utility to recover costs • Encourages innovation because utility can retain benefits longer
Cost-of-service rates	• Diminishes utility's benefits from innovation
Benefits allocated largely to customers	• Diminishes utility's incentive to innovate
Risk allocated largely to customers	• Increases utility's willingness to innovate • Unfair to customers if the utility captures most of the benefits • Creates a "moral hazard" situation
Ratemaking treats cost savings from conventional and new technologies the same	• Utility finds conventional technologies relatively more attractive
Book depreciation	• Can diminish utility's incentive to innovate • Can jeopardize utility's ability to recover the costs of existing assets
Prudence and "used and useful" tests	• Can deter utility from investing in relatively high-risk innovation • Protects customer against subpar utility performance or unexpected outcomes
Emphasis on reliability and safety	• Shifts interests away from cost-saving innovation
Favoritism toward certain innovations	• Jump-starts socially desirable innovation • Increases risk of choosing the wrong technology

Armed with a thorough analysis, companies will recognize the underpinnings of inertia. But the real challenge is not just the know-how, but closing the knowing-doing gap. Just as knowing how to lose weight is not the same as doing what is necessary to lose weight, understanding what is driving inertia is not the same as doing something to overcome it. Overcoming inertia calls for two leadership qualities—trust and courage—and it also requires the willingness to go against the grain, including all the factors that made the organization historically successful and all the qualities that were behind successful careers of many a leader. Breaking what is familiar and stepping into the unknown requires belief and trust in the rightness of the actions. It takes courage and grit to overcome fears—fear of failing, fear of losing credibility and reputation, and fear of letting teams down. Drawing on these traits, leaders must actively seek alignment and commitment from the organization using a combination of inspiration, logic, and conviction. The same traits are required for dealing with regulators who have to break from precedent and approve a risky innovation project. People tend to be criticized and penalized more for mistakes than they are rewarded for taking risks and doing what was different and risky.

Organizational change toward innovation begins by focusing on people. Leaders must identify the personnel who will support overcoming inertia and who will champion change efforts in the most critical tasks. It takes persuasion and influencing. The challenge that many leaders face is how long and how much persuasion is needed. Many leaders who have been successful in bringing change have shared that it helps to identify and group people into three main categories: (1) naturals, those who are die-hard supporters, (2) influencers, those who could be persuaded, and (3) naysayers, those who simply resist and block the change. Many leaders have confided that the smartest thing to do is to identify the naysayers and then ration one's time with them. Far too much time is spent in attempting to persuade those who won't be converted, rather than strengthening the case with believers and influencers. Another concern is organizations are inherently biased toward avoiding the sins of commission and not so much to avoid the sins of omissions. Leaders will proactively avoid taking actions that can fail over not doing anything and missing an opportunity. Accounting measures and KPIs track the performance of the decisions and initiatives that leaders take, not the opportunities that were missed. Over time, in a fast-changing environment, missing out on key trends can turn out to be catastrophic. If only Kodak had moved quickly into digital photography, for example, or Nokia and Blackberry innovated before Apple came in are questions that will haunt the leaders of these companies forever.

As we discussed in Chapter 2, the transition of scaled innovation into core operations is the place where most innovation fails because structural changes to legacy processes that don't align with the new processes are required, and these adjustments are counter to the established ways the organization has been working. Like an antibody that reacts to a foreign body, the strong forces of inertia easily overpower and reject innovation before the nascent process can gain grounding and support. With this in mind, it is incumbent on leaders to ensure that good innovation does not die prematurely and that sufficient structures are in place to prevent that outcome.

SECURING FUNDING AND INVESTMENTS

Innovation efforts in most power companies are funded by the core legacy business. This funding source creates an inherent tension between reaping immediate profits versus investing for the future. This tension is even more pronounced when companies have been in sustenance mode—meaning they've had steady revenue streams and predictable dividends and financial returns—for some time. To these companies, funding innovation is an immediate drain on profits. Pushing for capital investments with a higher depreciation rate changes the cycle for asset renewal. Where costs are borne by regulated utilities, returns above allowed regulated returns generated from new ventures flow back to the customers, hence leaving no incentives for shareholders to take higher-than-usual risks while not participating in the returns. If innovation has to rely on an additional rate base increase, it will be constrained by the affordability limits of the ratepayers (which are shaped by low- and moderate-income inequity considerations) unless there are policy mandates that force ratepayers to increase their affordability limits. For these reasons, leaders have an important role to play in securing funding and investments for innovation. Those who can secure funding more efficiently than others will determine which innovation programs are commissioned and executed. To effectively secure funds, leaders must focus on three areas:

1. **Understanding risks, uncertainty, and performance.**[5] Without a strong understanding of risks, uncertainty, and performance, a leader cannot right-size innovation investments, allocate funds, and make strategic choices.

2. **Persuading investors and regulators.** To support making riskier bets and provide a favorable regulatory regime to make investments, allow planned failures, and support long-term sustainability of the power sector.

3. **Conducting community outreach and garnering support.** To subscribe and adopt programs that allows for new distributed energy resources, such as rooftop solar PV cells, energy storage, demand-side management, and demand-response programs that creates new value streams.

This calls for leaders to rethink regulatory and investor relations functions. If there is a change in the risk profile of the industry, given VUCA conditions and the elevated levels of risks in the business portfolio going forward, it may be time to seek a different investor base and new ways for regulatory engagement. Such investor evolution is already evident as leading global utilities are floating green bonds and financing linked to environmental, social, and governance criteria that require new investor criteria. However, a vast majority of utility leaders feel stuck in their investor base and find their regulatory disposition to be fixed, that does not foster innovation. But as reality changes, holding on to the old ways might not be beneficial, and that applies to holding on to the legacy investor base and not leading the charge for a renewal of the regulatory practices. This notion may sound radical to many, but ignoring the signals for change can be a death march for the company and its investors.

INVESTING IN DIGITAL PLATFORMS AND ECOSYSTEMS

In today's digital age, leaders must recognize the relevance of digital platforms. They are not expected to understand the technical details or every new technology that is developed, but it is important that leaders are aware of the power of data and algorithms in driving business performance. Digital platforms present unprecedented ways to enrich customer experience, drive operational efficiency, and create new business opportunities. Technology breakthroughs like cloud computing, sensor technology, and AI/ML are at the center of development that yields direct benefits such as deeper understanding of and personalization of customer services. Engaging with customers, as we discuss in Chapter 5, is a fundamental shift in how traditional power companies use digital channels. With digital platforms, companies can create an internal information highway with systems of record that aggregates data from multiple sources, often termed as single source of truth. By bundling appropriate business processes and rules, companies can create an enterprise backbone that serves new business. With data collated and aggregated from across the organization and made readily available, companies can diagnose issues and apply corrective actions at a much lower costs on digital platforms. Companies like GE and Iberdrola have developed remote monitoring and diagnostic centers for power plants and large fleets of wind farms. Using the data in an AI/ML-based model, new value capture pools are unlocked in asset optimization, network operations, and trading and market operations. Siloed functions can now be integrated, which captures value that was not observable before.

Another way digital platforms are useful in future business is through their ability to create brand-new business models in the industry. It is inevitable that digital platforms will power new businesses related to DER-virtual power plants, blockchain-based peer-to-peer transactions, dynamic EV charging management, and demand response. Industry leaders who have a strong innovation agenda have already set up digital platforms to focus on these new business models.

Leaders must keep in mind four technological drivers to comprehend the business implications of engaging with digital platforms:

1. **Cloud computing.** An enterprise digital operating platform is at the foundation for innovating with data and digital applications.

To ensure this platform, digital leaders like Enel and Iberdrola are migrating their enterprise IT applications to the cloud. The plan is to create an operational data platform that allows faster, cheaper, and easier access to data to create services and products that generate new revenue streams. The cloud provides the underlying computing infrastructure with an on-demand, pay-for-use model, with flexing capacity for changing needs automatically. Using the cloud removes the need for utilities to invest capital and operation and maintenance time on an infrastructure of networks, servers, processors, storage devices, and applications. Cloud providers draw enormous scale advantages by pooling resources and thereby providing services at a reduced costs, often in the range of 30–40 percent lower. Cloud service providers are constantly developing offerings that go beyond computing and storage infrastructure to providing deployment platforms (or platform as a service) for system management, development, and test environments, as well as software applications (or software as a service) that rely on the use AI/ML.

2. **Sensors and communication.** The growth in sensor deployments has been propelled by the rapid cost reduction of computation and storage. Along with improved communication technologies, such as 5G, new opportunities for control and automation abound. Combined with cloud computing, sensors allow for new use cases, localized computation, and offer a greater choice for the creation of hierarchical control loops, such that local action in the field can be taken immediately while central control with the overall system visibility can take a wider area control action. Video streaming for surveillance and field monitoring, LIDAR[6] and satellite data for vegetation management and transmission line routing, phasor measurement units, and wide area monitoring and control for T&D networks are all technologies that can both sustain improvements and lead to breakthrough business models.

3. **Artificial intelligence.** With the growth in artificial intelligence and related disciplines such as machine learning, deep learning, and neural networks, new frontiers for innovation are opening up. When Google's DeepMind demonstrated to National Grid

how artificial intelligence can use computational models instead of domain expertise–based physics models, the industry saw the promise of AI/ML. The AI/ML field is developing; leaders will need to become knowledgeable regarding the business implications of AI as machines become sentient and more powerful at lower costs. DeepMind's solution was not adopted[7] by National Grid because of nonworkable commercial terms and conditions, but the technology serves a harbinger of what AI/ML can do in the industry—across the value chain from generation and T&D to customer experiences—in the future.

4. **Ecosystems.** In the digital world, it is often said that companies compete on ecosystem. The most significant shift that the power industry must adapt to is the increased reliance on third parties and platforms-based businesses and ecosystems. Such reliance is new for legacy leaders, who previously were not required to manage operations with a system of partners. Much can be learned from successful ecosystems like Apple and Uber, which have created businesses using large network effects and have attracted a large user base quickly, but the sobering fact is that it is not easy to build a profitable ecosystem. Hundreds of companies have failed to create an ecosystem, and their platform businesses never took off. Developing an ecosystem does not happen by chance or automatically. Orchestrating all these functional components is hard, and as research shows, a large percentage of ecosystems fail, even within the technology sector.[8] Leaders must drive clarity on what, where, and how they want their organizations to participate in the ecosystem. Should they be platform owners and be the primary orchestrator that builds and sustains the platform? Four central functions will matter in this area:
    - Attracting ecosystem participants and forging partnerships,
    - Providing architectural and technological leadership,
    - Designing incentives that work for all, and
    - Ensuring good governance and coordination.

Platforms that fail to create viable and thriving ecosystems fall short in at least one of these four functions. Platform-based companies often assume ecosystem participants will come based on the value proposition and the promise of the network effects. However, just like any other partnership, if the value allocation such as revenue sharing, end-market customer access, and incentives is not aligned or attractive, then platforms will not attract key players and will end up failing.

Based on observations of early industry ecosystem players (for example, GE's Predix, Siemens' MindSphere, Hitachi ABB's Ability, and Schneider's EcoStruxure), building and scaling a platform is highly complex, and platform owners must overcome many practical barriers to realize the promises of disruption that new technology can offer in this industry. Ecosystem platforms introduce new and diverse sets of competition and cooperation options. Two companies can compete in one scenario and partner in another. For example, Microsoft can be a cloud infrastructure and enterprise software provider to a power company and at the same time compete with the power company on a corporate power purchase agreement for utility-scale renewable resources to sell electricity. Hence, as often misunderstood, companies don't have to be platform owners all the time.

Digital platforms are shifting value propositions from quantitative to qualitative attributes where aggregation, connectivity and representation are becoming key indicators for success.[9] Aggregation of distributed resources is allowing new business models for customers, providing them with a choice to participate in aggregation and dispatchable clusters. Connectivity enables use cases like "connected cars" and "connected homes" that are selling value-added services beyond selling kilowatt-hours. Representation of customer data allows the identification of new consumption patterns, which in turn drive demand management and pricing programs. Data becomes the key fuel in the digital world (see Chapter 6).

Given the strategic importance of digital platforms and the explosive growth and success of digital platform-based companies, leaders cannot leave digital and ecosystem decisions to IT or specific functions within the business. Orchestrating all of these functional components is hard, and as research shows, a large percentage of ecosystems fail, but for those that succeed, the prize is huge.[10] The leaders who will succeed are the ones who are digitally fluent and can connect these technical developments with their businesses.

BUILDING A CULTURE OF INNOVATION

The quality of innovation decisions can be assessed on two primary criteria: (1) the strength of the idea and (2) the organization's ability to execute on the idea to deliver business value. The best ideas are the ones that promise value and that can be brought to market with the fewest internal and external hurdles. Across the globe, power companies are increasingly investing in new innovation facilities, creating innovation hubs in key locations, and investing in venture funds to seed and scale early-stage technology at innovation centers in, for example, California's Silicon Valley, Berlin/Munich, Cambridge/Oxford, and Shanghai. It is, however, too soon to be certain how these efforts will shift the century-old DNA of the power industry.

Having a "culture of innovation" is often considered the utmost priority by leaders; business executives tend to cite company culture as the main barrier when employees do not take initiatives or risks, or when they show a lack of customer service orientation. Such a diagnosis may be accurate, albeit superficial. What is more problematic is when the company neglects to address important underpinning factors regarding incentives, job security, and "return on politics" that shape organizational behavior and mindsets. When business leaders label a knotty organizational issue as company culture, it is often a cover for the reluctance of someone in the organization to take a decision. For this reason, this chapter began by discussing structure before culture. But culture change is not to be dismissed. Engineering a culture change entails a change in behavior, attitude, and norms; a new culture cannot be created in a short period of time, nor can it be changed on a dime. But when the new structures become the norm and are institutionalized into the organization's DNA, the old culture gives in to the new. And that must be the goal of every visionary leader.

SUMMARY

To thrive or even survive in a rapidly changing future, the power industry must turn to innovation. But industry leaders face great challenges in bringing innovation to their organizations. Not only do they confront today's VUCA conditions, but innovation needs to be developed and operationalized at a much quicker pace than previously. In the past, the power industry experienced steady growth in which moderate investments and incremental improvements were appropriate. Projects could be unhurriedly conceived, tested, vetted, and de-risked before commercial operation. But now and in the future, power companies will need to construct and manage a portfolio of risky ventures, moving from an investment practice that had sought only those projects that had a high likelihood of success to having to plan for many projects that may fail in lieu of a handful that succeed. To support this new way of operating and innovating, companies will need to reorganize with more-responsive communication and collaboration structures, and provide clear performance measures with the right incentives to the people. Leaders will likely face resistance and organizational inertia; and strong leadership will be required to ensure that innovation survives within the folds of legacy organizations. Additionally, for many investor-owned companies, the legacy investor base may not be the right fit for these innovation needs, in which case leaders must work with the board and investor relations team to gradually shift the investor base. Finally, innovation is not going to be a "one and done" matter; leaders must continually champion cultural and structural changes within their organizations so that innovation becomes a part of the organization's DNA going forward.

NOTES

1 Clayton M. Christensen, *The Innovator's Dilemma: When New Technologies Cause Great Firms to Fail*. 1st edition (Harvard Business Review Press, May 1, 1997).

2 The Department of Energy (DOE) has a list of the activities of incubators and accelerators: Office of Energy Efficiency & Renewable Energy, "Incubators and Accelerators," *Buildings*, https://www.energy.gov/eere/buildings/incubators-and-accelerators. Accessed September 13, 2021.

3 George Stigler, "The Theory of Economic Regulation," *Bell Journal of Economics and Management Science* 2 (1971), 3–21.

4 Edison Electric Institute (EEI), *Disruptive Challenges: Financial Implications and Strategic Responses to a Changing Retail Electric Business*, 2013, https://www.documentcloud.org/documents/1020804-disruptivechallenges.

5 Rahul Kapoor and Thomas Klueter, "Innovation's Uncertainty Factor," *MIT Sloan Management Review*, July 2020, https://sloanreview.mit.edu/article/innovations-uncertainty-factor/.

6 Lidar is derived from light detection and ranging, a technique for determining the distance to an object by transmitting a laser beam, usually from an airplane, at the object and measuring the time the light takes to return to the transmitter.

7 Sam Shead, "Google DeepMind's Talks with National Grid Are Over," *Forbes*, March 6, 2019, https://www.forbes.com/sites/samshead/2019/03/06/google-deepminds-talks-with-national-grid-are-over/?sh=2a7a28351914.

8 Ulrich Pidun, Martin Reeves, and Maximililan Schüssler, "Why Do Most Business Ecosystems Fail?" Boston Consulting Group, June 22, 2020, https://www.bcg.com/en-us/publications/2020/why-do-most-business-ecosystems-fail.

9 Ron Adner, Phanish Puranam, and Feng Zhu, "What Is Different About Digital Strategy? From Quantitative to Qualitative Change," *Strategy Science* 4, no. 4 (December 2019), 253–261, https://pubsonline.informs.org/doi/10.1287/stsc.2019.0099.

10 Pidun, Reeves, and Schüssler, "Why Do Most Business Ecosystems Fail?"

CHAPTER FIVE

Customer Engagement

What Are the Customer Engagement Requirements for the Future?

DOI: 10.4324/9781003353997-8

Customer engagement is the aspect of the power sector experiencing the most dramatic changes today; government and industry have become aware of this fact the hard way. In the US, for example, during the first wave of large-scale advanced metering infrastructure (AMI) deployment, between 2007 and 2010, gaps in industry understanding of customers led to significant problems: utilities did not foresee how serious and disruptive customer backlash would be regarding customers' safety concerns over smart meters and radio wave radiation.[1] Around the same time, in the UK, Prime Minister Gordon Brown's government, which had committed to a smart metering program, had to backtrack because of customer satisfaction issues. In the UK, the problem was not safety but the fact that meters were not compatible between different retail providers, meaning customers' ability to switch providers became almost impossible.

Such debacles—which have also occurred in other countries—are not only costly, but they also attract negative publicity and generate customer skepticism, causing further delays and cost increases. In an age of social media amplifying fear and discontent (and even misinformation), companies can be caught off guard by customer reactions. Leaving aside the broader sociological dynamics at play, it's clear that leaders must focus on meeting and anticipating customers' needs and wants.

Focusing on customer engagement is a radical shift for the regulated utilities industry. Traditionally structured as natural monopolies and serving a captive set of customers with guaranteed rate payments, utilities are now subject to customer choice, adoption, and acceptance. New customer-responsive initiatives such as demand-side management, greater energy efficiency, and electric mobility (transportation) hold promise for strategic growth. In addition, new sources of energy— rooftop solar PVs, battery storage systems, community generation, and microgrids—are key to satisfying growing customer interest in the power industry meeting sustainability goals. And because upgrades to support this additional operational infrastructure (while maintaining the existing grid) will depend on ratepayers approving capital costs, positive customer perceptions are important.

DRIVE CUSTOMER CENTRICITY FROM THE TOP

E very company approaches customer engagement in its own way. Some leaders focus on relationships; others seek to secure revenues. How leaders engage with customers varies based on the role they see themselves playing in driving the success of their businesses. Some conduct social visits with strategic high-valued customers important to the business primarily to foster relationships and renew trust. Some are merely transactional and take a hands-off approach to customer matters, or leave customer interactions to the marketing or the customer services team. Some are more hands-on and become involved more directly in detailed business discussions, while others may stay away from direct interactions until a big transaction is executed. Historically, how senior business leaders involved themselves in customer engagement did not matter much, but this changing trend in customer expectations means leaders will need to adjust to fulfill the customer engagement role. When customers demand that the CEO be the brand ambassador, the face of the company, and the company's vocal champion, then the board and senior leaders must take a hard look at the abilities of the C-suite. What might be a nonissue in the traditional legacy organization could prove to be damaging to the company's success if not handled correctly today.

In short, it will be increasingly difficult for leaders to do their jobs without a solid understanding of the end-to-end customer experience. The wide variation in customer expectations calls for deeper awareness of what customers are expecting from their power companies, and which interventions they seek under different situations. Understanding customers is a tricky business, as anyone in a business-to-consumer business will attest. Although metrics and tools—such as the J.D. Power score or net promoter scores—are available, these measures do not get to the root of what drives customer satisfaction and loyalty. Tools and techniques such as surveying, focus groups, persona analyses, and journey mapping can provide a snapshot of certain needs and expectations; however, these tools do not synthesize customer goals, desires, and willingness to pay in a consistent and systematic manner, leaving much room for interpretation and intervention. It is only in recent years that companies such as Amazon, Uber, Netflix, and Airbnb have shown how digital platforms and innovations with customer data can provide deeper understanding about the individual customer that can be converted into customer delight and loyalty. Companies that obsess over their customers' experiences do so overtly and take actions to bolster customer service in every way possible. Customers and investors see and experience these services and reward such businesses with their wallets and valuations.

Customer-centric leaders know that customers develop new needs, wants, tastes, and expectations not by accident but by the types of service and the level of service they are receiving in other arenas. What was once a novelty quickly becomes a table stakes expectation and demand. The presence of a service or benefit does not uplift customer delight as much as the absence of a service or benefit diminishes and reduces the company's value. For example, Starbucks, an American coffee shop chain, took the commodity of coffee and elevated it by creating a customer experience for which the market is ready to pay a premium several times over what a cup of coffee costs. Similarly, over the past two decades, we have seen power industry customers shift from focusing on price, reliability, and quality of service to focusing on service options, control, and user experience.

In recent power outage events in the US, UK, and Europe resulted in public outcry regarding the lack of clear and timely communication about service restoration. Few people saw the distinction between the uproar over the loss of power and the poor communication that followed. But the fact that there was no information related to the outages, why it happened, when power would be restored, and whether such an event would happen again angered customers as much as the actual outage.

Understanding customers is pegged to two other priority areas: innovation (discussed in Chapter 4) and data management (discussed in Chapter 6). Analyzing customer data for behavior patterns provides a deeper understanding into what matters to customers. Data are used both for diagnosing issues and then for addressing customer needs by creating new products and services that have better chances of driving business growth and customer loyalty. For example, many companies are investing in new customer channels to provide customers with multiple options for self-servicing bill payments. Further, understanding customers is not a one-shot affair; it is an ongoing process for which leaders must take a lifecycle view of customer engagement. Using a lifecycle view or customer journey mapping aggregates customer experiences starting from the point when a customer enters into a relationship with the company until that relationship is terminated. It encompasses all the customer interactions— account services, receiving and reacting to notifications and alerts, commercial transactions, and the acceptance of new plans or services. Understanding happens not just at formal touch points but through all the data that the customer reveals—even informally through social media channels, for instance.

INVEST IN THE SIX PILLARS OF CUSTOMER ENGAGEMENT

Leaders need to push their organization to look for ways to observe, learn about, and reflect on customer experience and what the organization can do to improve it. This sum total of all experiences— which can be highly subjective from the customers' perspectives—makes it difficult to find a good measure for customer engagement. Measuring customer service levels is itself tricky, but it is important to track progress by constructing a reasonable set of performance measures.

Despite the speed at which customer expectations are changing, leaders need to take measured and calculated steps to transform their organizations to focus on the customer experience. Leaders must also consider whether they are settling for mere incremental change when bold moves are necessary. Six pillars support a customer-focused power company, and leaders who move quickly to develop these areas will be better positioned for building their company's future:

1. **Fit with total customer experience.** When companies focus on total customers experience, they gain clarity on any tradeoffs that are made. As Amazon's founder has often stated that given the choice of obsessing over competitors or obsessing over customers, he always thinks about obsessing over customers. Technology and digitized processes provide enormous advantages to observe the end-to-end experience of the customer and to take actions, including inventing ways to serve. In recent years, investments in omnichannel customer experience and having 24/7 availability to communicate and transact business have been rewarded by explosive growth and acceptance. Observing and drawing customer sentiments from social media platforms such as WhatsApp, and voice-activated devices (e.g., Amazon's Alexa) for interaction with businesses has become common. The use of Twitter and social media platforms to broadcast messages has increased. As technology makes observation and understanding cheaper, easier, and better, leaders must not lose sight that technology in all its forms is simply an enabling agent. The real expertise is in inventing new ways to elevate the customer experience beyond what the customer knows or can articulate.

2. **Personalization.** With advances in data and analytics, it is possible to track and understand an individual customer at scale. This ability has unfolded the possibility of personalized services for each and every customer. Personalization addresses the uniqueness of what, when, where, and how each customer engages with the company. Companies can personalize customer experience by giving special offers, allowing distinctive channel experiences, and/or providing special service terms and product customization. Personalization draws on contextual knowledge from a vast array of sources to reduce friction and provide efficient services to the customer. Customer knowledge is used to steer purchasing behavior and shape customer needs. Personalization has been made possible by advances in new digital technologies coupled with data analytics powered by artificial intelligence, which engender a high degree of personalization at scale and at affordable costs in a way that was not possible before. Furthermore, the adoption of automation

and simulation tools can expedite personalization in customer engagement and increase its effectiveness.

3. **Mobile-first option.** The explosive penetration of smartphones has created a mobile-first culture in most parts of the world. Digital solutions go beyond the web portal to a mobile-first environment. Because of mobile phones' flexibility and convenience, mobile solutions offer a rich experience. The mobile-first view serves as the nucleus around which additional choices and control are provided. Examples include interactive voice response, mobile apps, smart home devices, chatbots, and text messages. Alerts and notifications, especially those pushed by the utilities using native mobile messaging services[2] provided by telecommunication providers are adding to the experience. The ability to switch between different channels requires that the customer experience is seamless, such that customers can pick up on a computer where they have left off in the mobile interface because there is a common source of data and integrated back office processes. This common data source enables seamless transfer from the automated systems to customer agents, reducing friction and common customer pain points. Many power companies have started building an ecosystem of partners, such as financial services, that can extend their offerings to new utility businesses including e-mobility and DER.

4. **Prosumer support.** The growth in customer-owned distributed generation connected to the grid has created a class of *prosumers*, individuals who are in contractual agreements to buy and sell electricity—in other words, individuals who are consumers *and* producers. Interactions require active two-way communications between the utility or the network operator/DSOs and the prosumer on switching, controls, tariffs, ownership, and services. Prosumer engagement calls on cultivating mutual trust, and it requires the utility to see value through the eyes of the customer. This approach is new for many network service operators. Energy conservation, demand response, and energy-efficiency programs depend on the commercial fit with the customer's lifestyle and convenience that changes in real time. A household nursing a sick child might have

a different approach to demand management than a commercial building that is interested in optimizing consumption. Some customers may be passionate about energy conservation and may require detailed information on that topic. Others may take a "connect and forget" approach.

5. **Customer as a strategic growth lever.** Customer programs and services present new frontiers for industry growth decoupled from the historical reliance on sales based only on kilowatt-hours. Electrification of transportation, district and consumer heating, and building and home energy management are dependent on customer adoption. Depending on customer segments and buying patterns, power companies will need to introduce new products and services in the marketplace, meaning that customers will likely unlock new value streams vital for the longer-term strategic growth of these companies.

6. **Regulatory alignment.** Existing regulations limit the extent to which power companies can adopt new models of customer engagement. Changing customer demands and slowdowns of modernization efforts can pose a risk if companies are late to adapt and are constrained by regulations. Leaders who have taken a proactive stance and have been actively involved in shaping regulations in tune with customer needs and wants have achieved better success in customer engagement.

ADOPT A SYSTEMIC APPROACH

Customer engagement is a dynamic process of interactions and ever-strengthening connections with customers. It requires a methodology that reflects its changing nature. What companies can know about customer needs and wants is frequently not aligned with what companies do. Often, the challenges of understanding customers result in complexities in how organizational goals are set and actions and priorities are decided given that they may often contradict and

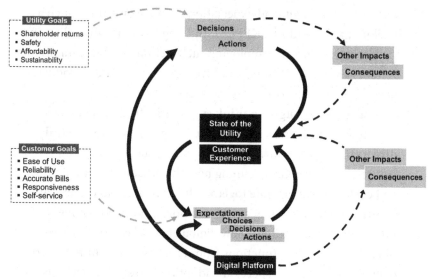

Figure 5.1 *Dynamic Nature of Utility-Customer Interactions*

even in direct conflict with each other. Even applying the traditional approach of customer segmentation can be difficult, given the fog of uncertainty looming in the environment. Technology can provide customer insights, and customer data and analytics can indeed support hyper-personalization. But regulated utilities cannot provide benefits to one group at the cost and dissatisfaction of another. Without proper safeguards, serving one constituent can create unintended inequity issues. To overcome such situations, a systemic approach that focuses on the various cause-and-effect threads and interrelationships can provide greater insight and clarity for decision-making. Such an approach must include customers as part of the sense-learn-act system (see Figure 5.1).

Utilities must take three steps to better integrate the customer into their overall operations and to activate proper customer engagement:

1. First, **utilities need to build customer awareness and reduce the gaps in customer knowledge and understanding.** By analyzing customer user data and the nature of interactions with advanced AI/ML, companies can gain deeper understanding of customer

response, sentiment, and expectations. Many companies, such as Exelon, Enel, Iberdrola, KEPCO, and Engie, along with utilities in California, are already employing such techniques. Both traditional vendors and new entrants are actively developing and introducing new software tools to accomplish these tasks.

2. Second, **utilities ought to think systematically about customer engagement processes and execute incrementally to carefully direct their course.** Bringing all the major stakeholders and entities together entails defining how the broader system of utility and customer will operate together. Incremental and piecemeal buildout such as building a new web portal, adding incremental information, and adding a new digital channel can help test the interplay of different functions from IT to capital planning to grid operations. But wider success in adoption and scaling depends on the interlocking of multiple stakeholder interests—from regulators, investors, utility boards, and executives to community groups and customers. Serving one constituent and not the other, or worse, serving one at the expense of the other, can result in public outcry, legal suits, and regulatory and investor backlash. These situations have already affected utility programs negatively, putting for instance, AMI, demand response, DER, and electric vehicles (EV) at risk.

3. Third, **utilities must build for flexibility, adaptability, and customer shaping.** Customer programs will need to continually adapt to changing trends, and companies will be modifying their customer program offerings and services ongoingly. These activities require a strong foundation. Digital platforms provide a solid backbone of data and process integration of the foundational customer functions such as customer relationship management, enterprise resource planning, enterprise asset management, meter data management (MDM), and others. Once this foundation is in place, it allows flexibility and adaptability to support new and changing customer-related applications. Many customer groups often balk at building a foundation, because they think that it is time- and resource-consuming and that it does not provide

immediate, tangible values. But this notion is wrong. Digital customer experience platforms today allow easy and fast adaptation not only for traditional applications but also for EVs, smart homes, and sustainable cities. In addition to their flexibility and ability to be expanded, over time these platforms help in reducing total lifecycle costs.

Although many companies have embarked on these systemic customer experience steps, the vast majority have not. In most jurisdictions, utilities command a high level of trust on matters concerning electricity-related programs and appliance advice.[3] Adopting a systemic approach—and, in particular, working backward from what the customer needs and wants—will position the utility to chart its grid modernization journey with greater chances of success.

BUILDING CUSTOMER ENGAGEMENT AT THE CORE

To develop a strong customer engagement program, leaders must address two fundamental issues that have broader strategic implications: customer engagement maturity and selection of technology.

CUSTOMER ENGAGEMENT MATURITY

The customer engagement transformation process, however dramatic, must progress in carefully engineered steps. Once a power company has completed a reasonable assessment of where it is today in terms of customer engagement, it must design its course to a higher maturity stage—which may range from basic to a personalized experience. The projected maturity stage is determined by the company's strategic plan and must be aligned with its strategic vision of a sustainable future. This multistage view forms the basis for detailed plans and experimentation projects. Each rung of the maturity ladder depicted in Table 5.1 represents

Table 5.1 *Customer Maturity Levels and Attributes*

	Basic	Connected	Empowered	Personalized
Technology	Web portal–based	Multichannel experience platform—mobile, interactive voice response (IVR), mobile push notifications	Self-configuration options with a common system of record for customer-related matters • Notifications and alerts of customer choices • Limited statistical and clustering analyses	AI/ML-based analytics and insights platform
Engagement	Generic one-size-fits-all approach	• Basic customer awareness • Company decides where to send customers	Customers self-select transaction choices based on their needs and requirements	Company understands, personalizes, and dynamically suggests and nudges
Transactions	Common products/services are targeted to groups of customers	Customer-channeled to pre-assigned response system based on obtaining some information	• Customer can make choices in how and when to transact with the utility • Journey-based transaction with emphasis on "moments of high impact" • Extensive customer knowledge leads utility to customer-specific products	• Personalized transactions based on patterns mapped automatically to individual's financial, environmental, and technology preferences • Proactive nudges and informed guidance to suit customer needs

	Basic	Connected	Empowered	Personalized
Collaboration	Customers adjust and adapt their needs to fit with services/products offered	Products/services developed with minimal customer interaction	• Deep knowledge of customer base leads to customized processes (e.g., systems/tools for interaction) • Utility works hand in hand with customer to create tailored product offering	• Partnership with customers to cultivate and shape interests and to attract attention to meet mutual needs
Education	Minimal and generic hotlines are usually limited to the service accounts	• Channels of interaction are dictated to customers • Conducts seminars on products and services offered to various client groups	• Customer is able to acquire information through multiple sources (e.g., web, phone, individual care) offered by the utility • Deep customer knowledge leads to pertinent product information being pushed out to targeted customers	• Personalized outreach and education messages using a range of segmentation attributes and interest areas (e.g., environment, innovation, support, and finance)
Impact	Utility focus is on revenue protection	• Utility focus is on connection with customers—as a foundation for improved engagement • Reduced first response and resolution times with integrated platform	• Utility explores new streams of revenue • Information and insights to make choices • Productivity improvements from self-service • Customer needs are captured systematically	• Customers are deeply interested and pay attention regularly to their utilities • Customers become a core source of value creation and capture

a level of competence achieved through additional investments in technology, tools, and process changes that build customer engagement.

Companies should limit their future outlook to three to five years, beyond which planning typically becomes too speculative for setting an action agenda. Table 5.1 shows maturity levels from basic to personalized, and provides description of the six salient attributes in each of these levels. Leaders must note that every situation is different and that an organization might have its own strategic motivations to set the maturity levels—such as aspirations to benchmark against industry leaders or to meet specific customer needs. The maturity ladder calls for periodic reviews and updates, and leaders must ensure that conditions remain valid for the maturity plan to hold.

The frequency of review will depend on how dynamic and volatile the environment is. Jurisdictions where a convergence of decentralized energy services, customer-owned generation, and deregulated electricity markets are underway will most likely require more frequent reviews and changes. In such cases, updates may be necessary on a monthly or quarterly basis. The maturity ladder is a living tool to define customer positioning for the company and then use that information to match short- and mid-term commitments that are in line with evolving customer expectations.

SELECTION OF TECHNOLOGY

Technology plays a defining role across all areas of customer engagement—from customer understanding, journey mapping, to enabling use cases. Given the strategic importance to the business, technology selection cannot be left to the IT department or the customer service department. In addition, with so many technology options at different stages of development, provided by both traditional companies and upstarts, technology selection is a daunting and complicated task. To provide a structure to the selection process, we introduce a simple framework to delineate and describe the four pathways to customer engagement in Figure 5.2:

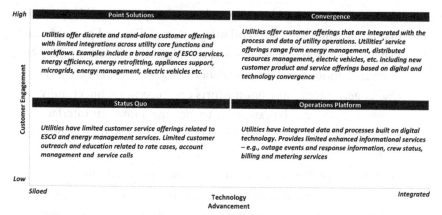

Figure 5.2 *Customer Evaluation Scenarios Framework*

The framework shown in Figure 5.2 relies on two driving factors: (1) the company's appetite for technological advancement and (2) the company's aspirations for customer engagement. The four quadrants range from status quo to convergence:

- **Status quo.** Companies in the "status quo" quadrant focus on incremental system and process upgrades to maintain or marginally improve current customer service levels. The portfolio of customer services is predominately designed to support traditional commodity supply with limited inroads into new use cases. "Status quo" utilities are mainly followers and strive for self-preservation and relevance in the industry with limited and low-risk investments. Customer engagement, therefore, focuses on maintaining general trust and avoiding negative perceptions that might affect business continuity, and create barriers in getting regulatory approval for investments.

- **Point solutions.** Companies that opt for "point solutions" identify new products and services as discrete and stand-alone solutions. Running multiple-point solutions can avoid cross-functional challenges and keeps the program under the control of one function or department. Companies in this quadrant have customers who are reasonably engaged, which allows the utility to make modest to high investments, grow, and sustain its base. Companies in this quadrant may defer wholesale back-end integration of data and processes and may launch discrete products and services. Over time, a proliferation of point solutions and discrete data management can become a challenge. System integration efforts can become architecturally complex, and it can be expensive to manage numerous interfaces and systems with various vintage versions, lifecycle management requirements, and hidden vulnerabilities.

- **Operations platform.** Companies in the "operations platform" quadrant are those that focus on investing in foundational data and process platforms, and on leveraging and extending the current operating platform to develop new customer capabilities. Examples are building analytics solutions on an AMI/MDM platform or enterprise platforms like SAP, Salesforce, or Oracle. If not planned, then the existing systems can become a constraint if new customer use cases and data needs cannot be met. If the customer changes are not dramatic, companies can balance customer engagement with the risks of rapid change.

- **Convergence.** Companies in the "convergence" quadrant rely on an integrated portfolio of customer-centric products and services to create strategic and long-term growth. They invest in digital platforms that integrate data, processes, and workflows. They strategically position themselves to take advantage of advances in integrated data backbone, analytics, and in artificial intelligence and machine learning, and they provide value-added services depending on customer needs. Companies in this quadrant actively seek new revenue streams by adopting a customer-back view on offerings to create. An integrated backbone does not necessarily mean constituting a large expansive program often termed as the big bang approach, rather a holistic approach to addressing the use cases. This requires a strong technology evaluation and adoption capability in the customer domain and savvy data management to navigate privacy and security issues.

MOBILIZING THE ORGANIZATION

A senior utility executive once remarked, "Customer engagement is so hard because any obvious improvement costs money for the company that customers (ratepayers) don't want to pay." This statement underscores why finding creative ways to address customer engagement is critical. Digital platforms hold a unique promise in that solutions do not have to be more expensive once the foundational components are in place. Digital channels allow shifting customers to lower-cost channels (live voice calls to self-serve portals and chatbots), thus saving expenses for the company. Multichannel enablement empowers customers to engage in active two-way interaction. If we examine the companies that are known for customer service, we find that their businesses depend on customer loyalty, and it is not by accident or coincidence that these companies have built a dedicated customer base. Customer loyalty is created at an emotional level founded on trust, commitment, and partnership and by providing reliable services when customers need them most—what's termed "high-impact touch points." Most power companies understand the importance of customer engagement, and many have established a chief customer officer. But to practice across the organization at various touch points on the front lines by the rank and file requires clarity and alignment on high-level intent and means to deliver through middle and back office support systems. Companies must enact six practices to ensure such a strategic partnership with customers:

1. **Understand customer aspirations.** Leaders must understand the situational context of their customers. One utility serving one of the poorest states in the US found out that customers did not want their bills to increase at all. Customers mobilized local politicians and regulators to delay their AMI investment for a few years. Customers even accepted a modest reduction in their reliability levels because lower bills were that important to them. Despite the fact that smart metering deployments countrywide were at their peak, the utility had to adapt to its customers' needs. Similarly, in

other situations, the demand for sustainable and cleaner resources is pushing companies to change their portfolios. Leaders must start with customer aspirations in deducing and deciding what their organizations should do.

2. **Purpose-built operations.** The company's operations, systems, processes, and organizational structure must be set up to make it easy to serve customers. If a service agent is unable to provide an estimate on a job's cost or duration; when there is no line of sight to meeting customer service demand; or when customer needs and desires are not captured well and do not guide innovation decisions, then it is not easy for the operations to serve customers. In a changing environment, the need to sense and pivot to different customer preferences or requests is even greater. No perfect operation design suits all situations, but when leaders build operations purposefully to meet customer needs, the chances of success are greater.

3. **Manage costs to serve.** To pivot to an innovative and customer-centric enterprise, companies must support customer aspirations with real investments. Justifying these investments is not always straightforward; it is hard to know the precise linkage of customer satisfaction to financial outcomes. Because customers bear the investments and operations and maintenance costs in regulated utilities, which include costs incurred in customer outreach, education, and messaging, the amount spent by utilities needs to be supported with a business case. After a European utility acquired a US company, one US customer remarked, "I don't understand how a foreign retailer who had to set up a new business could provide a cheaper rate than the retailer who had common roots with the utility that served this area for many decades." The answer is that the new retailer must have a better cost-to-serve model than the incumbent as a result of thoughtful investments in building a foundation that provides a lower cost to serve.

4. **Manage for performance.** Customer engagement programs must deliver on desired outcomes. Peter Drucker wrote about "management by objectives"[4] several decades ago,

and more recently, companies like Google have popularized this concept. Without having appropriate measures that are defined, monitored, and tracked, management is handicapped in achieving goals. Measures, both quantitative and qualitative, allow the program to stay on course and, should it fall off, allow it to get back on track by making necessary changes and adaptations. A weak or missing performance measurement—or monitoring and tracking system—will increase the likelihood of leaders becoming misaligned with others in the organization, resulting in churn and wastage, and even worse, adding costly workarounds that reduce overall effectiveness. Companies fall into a trap by overdoing customer surveys and focus groups, and then employing technology that does not deliver on customer satisfaction or create a richer experience, because they do not employ measures that matter, nor manage based on relevant data. Because measures related to customer experience often tend to be subjective, noisy, and fuzzy, it so critical that measures are crafted to serve specific purposes, and that a system of feedback loops informs necessary actions.

5. **Pay heed to legacy constraints.** Many power companies are reinventing their customer engagement from a legacy position. Pivoting from a legacy position requires knowledge of the current business to determine the course of change that will stick and at the same time not upset the current business to premature destruction—a tough task. The global electrical and electronics giant Philips adopted strategies to become a differentiated player and to claim premium pricing in home lighting solutions that can now be tailored to customer tastes and preferences. Light bulbs before this move were a commodity product, and many large companies like GE divested their businesses in this area. While Philips was pivoting to its new Hue business, the company gradually exited the legacy business models for light bulbs and invested in cloud and IoT to provide a complete customer experience with different choices in colors and configurations to suit the customer's mood, feelings, and preferences.

Similarly, automobile manufacturers that have a dual strategy anchor different sets of customer experiences to different customers. Volkswagen AG has the VW brand as well as Audi and Porsche, Toyota has its Toyota Camry and Lexus. The customer experiences for Audi and Lexus are different from Camry and VW. To break from legacy practices, utilities have to deliver on customers' aspirations and limit costs to what customers are willing to bear. Lessons from legacy companies that have made this shift to deeper customer engagement—whether it is Burberry with new digital channels or Unilever with sustainability in its branding, or Nestle with its pivot to healthier food products—all suggest that these shifts have been achieved through multiyear enterprise commitments and persistent actions in the face uncertain outcomes. These companies employed a variety of models to steer the legacy business to the new. To do so successfully, companies must begin with the knowledge of legacy constraints and the barriers that need to be systematically broken down.

6. **Take customer privacy seriously.** Imagine the company has a customer profile that is constantly updated based on artificial intelligence and machine learning routines. Customer service agents provide personalized pricing plans, advising based on consumption patterns, and recommending products that fit the customer profile. The information that companies would like to gather from their customers and what these companies are willing to make available to other parties in the ecosystem create an ongoing tension that is likely going to increase. A range of customer behavior patterns derived from these data can be used for purposes outside the power services domain and advertising (for example, data volume, data types, information classes, and modes of sharing). A key decision factor is determining which data will be shared and how the data will be used and by whom (discussed next, in Chapter 6, "Data Management"). Using sophisticated tools that trespass on customers' private lives without their knowledge and consent raises concerns. Recent data collection and sharing regulations such as the EU's General Data Protection Regulation have illustrated the complexities

in compliance for a variety of businesses. The degree and nature of concerns across customer groups vary. Leaders have to be vigilant and ensure that their organizations comply to protect themselves and their customers from the possible adverse effects of privacy infringements.

SUMMARY

In contrast with past models, customers will play a defining role in determining the success of power companies in the years to come. To craft a robust and effective customer engagement program, power companies must develop a systemic understanding of the customer and orient their operations toward serving the customer—actions that mark a major shift for the industry. Traditional relationships between customers and utilities mainly concerned account management, service calls, and billing-focused interactions, but under the current wave of change, customer relationships will be radically transformed not only to meet evolving expectations but also to secure future growth. This transformation will be a journey with thoughtful steps taken toward maturity and technology enablement. No two companies' journeys will be the same, and successful navigation will require leadership at all levels. By studying customer data using digital platforms and analytics, companies can form an innovation agenda for new products and services that can be brought to market at unprecedented speed. Although this undertaking is complex, with many interlocking functionalities and disciplines, a handful of practices can ensure leaders creating effective decision models and playbooks.

NOTES

1 American Cancer Society, "Smart Meters," September 24, 2014, https://www. cancer.org/cancer/cancer-causes/radiation-exposure/smart-meters.html.

2 Native mobile messaging uses the cellphone's mobile network and not the Internet.

3 Smart Energy Consumer Collaborative, *Data Analytics: Unlocking the Consumer Benefits Report* (2018), https://smartenergycc.org/data-analytics-unlocking-the-consumer-benefits-report.

4 Peter F. Drucker, *The Practice of Management* (New York: Harper, 1954).

CHAPTER SIX

Data Management
What Are the Data Management Requirements?

DOI: 10.4324/9781003353997-9

"The World's Most Valuable Resource Is No Longer Oil, but Data," *The Economist* wrote in 2017,[1] building on British mathematician Clive Humbly's oft-repeated 2006 pronouncement that "data is the new oil." What did Humbly and *The Economist* mean when they compared oil to data? Is data as valuable a commodity as oil? Today, the parallel between the potential value of large reserves of oil and of large reserves of data, both unextracted, is much clearer. As the promise of algorithms and AI/ML-powered analytics unfolds, owning large data sets equates to future value. For the past decades, as utilities have embraced digital technologies, the looming question has been how to use all the data that gets generated. This trend became more pronounced after smart meters were installed, and more lately when sensors were deployed in field devices. While we find that many managers grapple with this question of what the appropriate use cases are, it is worth reminding that this question is not new. In a different context, the question was raised when digital relays were introduced more than thirty years ago. Therefore, the fundamental approach that we advocate in this chapter is not to sort out how to use the data, but to focus on the business and functional outcomes that need to be driven from the data that is generated. Working backward from the business outcomes and decisions that need to be realized, companies need to determine the data that must be collected, aggregated, and analyzed.

In the current context, as the first wave of deployments of advanced asset management, workforce optimization, and system planning have shown—data is key in supporting decision-making and in creating value streams for customers and operations. Data management—including understanding how data represents physical phenomena, how the aggregation of disparate and disjointed data sets can offer new insights, and how data connectivity and collaboration can create new workflows and value streams—is a strategic competence necessary in the drive toward a sustainable future. Data management is no longer simply an IT issue to be dealt with by the CIO; it is at the center of business positioning and requires the attention of the chief human resources officer, the CFO, and others involved in business operations.

Unlike oil (and this is where data departs from the "data is the new oil" analogy), digital data is what economists call a "nonexcludable good"—that is, a good for which one person's use of the good does not exclude its use by another person. Unfortunately, the ability to duplicate, transmit, store, and retrieve data with almost no marginal costs is accompanied by issues of privacy, security, and data persistence. In addition, the data that is advantageous to one company may not be so for others, and the data that is advantageous at one point in time may not be of any consequence at another point in time. Despite this range of challenges, data's enormous potential value makes overcoming such hurdles worthwhile.

We have discussed that it is a leader's job to chart their company's course to the future. Data serves as the evidence and the empirical guide for leaders as they imagine the future and define the company's North Star; it keeps decision-making within the bounds of reality and constraints. Data management is the central pillar, indeed, on which much of the future applications for operations, customer experience, and managerial decision-making will stand. Hence, leaders must prioritize in ensuring that their organizations' data management capability matches to the needs of the operating environment and market context. This chapter delves into key data management areas where business leaders must focus.

KNOW WHAT IS POSSIBLE

As digital platforms become ubiquitous, digitization and data have become central to shaping and controlling business processes and analytics, providing important insights for decision-making. Yet, how to tap the full extent of the value has been a challenge ask. A CEO of a large utility once expressed his growing concern with data, stating, "We have all [this] data generated from new digital systems, but we don't know how to monetize it." Many leaders share this frustration, understanding intuitively that much can be done with data, but struggling to find the means to extract value from it.

Most leaders have lived through the consequences of decisions based on poor-quality data, and they understand that not trusting data (or not having trustworthy data) results in lack of confidence in decisions based on that data. In addition to addressing quality issues and investing in data quality, leaders must know how to properly use data, which can be unwieldy, inscrutable, and complex. When asked what is possible with data, leaders' focus is usually limited to technical aspects. *Is the data trustworthy? Is there a master data management plan? Is there a system of records? And is it the single source of truth? How secure is access? How much manipulation is needed to make it useful?* These are foundational technical questions that data architects must address. However, to use data to support their business, leaders must identify the innovations that can be created from data that will add value to their companies. To benefit from their data, leaders must make innovation a strategic priority (as discussed in Chapter 4); simultaneously, they need to systematically probe specific data management attributes that impact specific business and corporate strategic outcomes.

To know what data can do for a business, we need to understand the data that is accessible and available (illustrated in Figure 6.1). This understanding means being able to evaluate the sources, the ownership and exchange of data, and the communication systems used to transfer data from the source to the point, where it can be used with other data sets for insights and decision-making. To determine how much value can be captured and realized, data usage and its contribution to margins and risks must be assessed. Consider, for example, the electricity usage data of a residential customer that is read by a smart meter and stored in an enterprise system. The data reflects the activities of the customer household. Depending on the jurisdiction, meters that measure the usage may be owned by the customer, utility, a retailer, or a metering company. Data may be transmitted by a private or public network that may be owned by or leased from third parties. Finally, the data is collected in back-office systems of the utility and may be stored in a public cloud. Along these lines, other parties providing additional services may also be

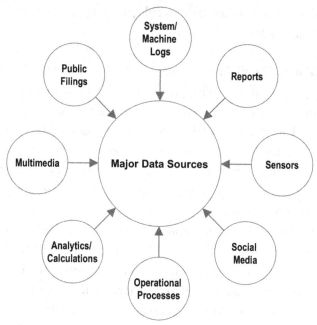

Figure 6.1 *Major Data Sources*

involved in the value chain. Hence, if we ask who owns the data, we can get many answers:

- The customer who uses and pays for the electricity (the data reflects customer activity),
- The utility/retailer or metering company that gathers and collects the data,
- The original equipment manufacturer, software provider, or cloud provider that works with the data, and
- The commons, because everyone can benefit from the data.

Ultimately, ownership of data depends on policy and business negotiation; it is a leader's job to ensure that their organization has a data strategy with a stated position backed by a business case to articulate data ownership, access, analysis, and usage. For these purposes, it is useful to group data sources by the natural boundaries of functions and ownership in the power business value chain: retail data, network data, and market data (see Figure 6.2):

- **Retail data.** Retail data consists of metering data related to the consumption of electricity, production of behind-the-meter resources, and customer account information and location.
- **Network data.** Network data consists of historic and real-time grid operation data containing operating parameters like voltage, current, frequency, real and reactive power, power quality metrics, grid configuration, asset status, planning and maintenance data, and outage information.
- **Market data.** Market data consists of data generated from different market operations such as weather data, spot market data, load data, generator data, unit level production and consumption (historical and real-time) data, and flexible resource data (e.g., energy storage, fast-ramping generators, and demand response data).

Data is exchanged between systems and analyzed and used in systems other than where it was sourced. This exchange can be a one-way or a two-way transfer. Power companies have decades of experience with data exchange among various entities, both internal and external (for example, regulators, balancing authorities, and customers). For exchange between heterogenous systems—including solutions from different vendors and multiple devices and entities—companies typically select and implement using existing protocols and standards for data exchange. These standards are not always optimal for a particular company setup, and it

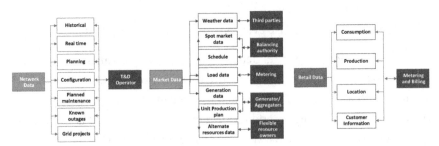

Figure 6.2 *Categorization of Data Sources*

may take years to develop and ratify standards. In addition, standards often leave room for interpretation, which creates the need for elaborate interoperability testing to confirm that the systems can interconnect and operate as desired. Many of the data exchange standards are highly complex and deal with multiple abstraction layers[2] that can only be engineered by specialized personnel. Despite these challenges, companies are wise to adopt standards and protocols—that provides companies with a larger vendor base, better commercial terms, and a larger pool of skilled personnel than they would have, were they to adopt proprietary protocols (Figure 6.3).

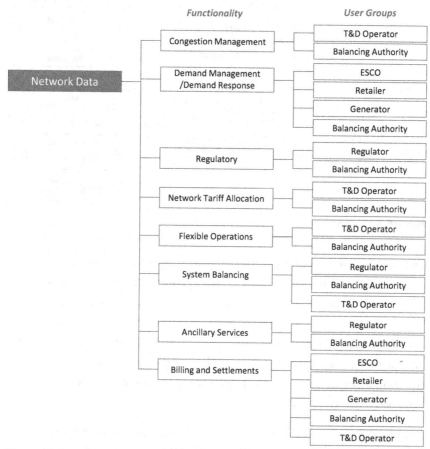

Figure 6.3 *Example Mapping of Data to Functional Use and Users (Non-Exhaustive)*

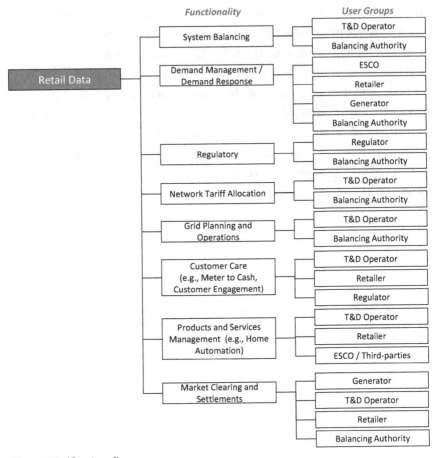

Figure 6.3 (Continued)

Data can be used for many things, including operational efficiency, worker productivity, real-time grid management, customer behavior analyses, and market research. New data-driven use cases are associated with resilient microgrids, electric vehicles, improved safety, and asset utilization. To prioritize how they will employ data, leaders must articulate what use cases are of real value for the organization based on the company's strategic roadmap. Figure 6.3 maps data according to functional use and users.

Because data items can have multiple users and uses, leaders must comprehend the full range of data functionalities and must know who will use the data so that they can ascertain the data's value and determine the

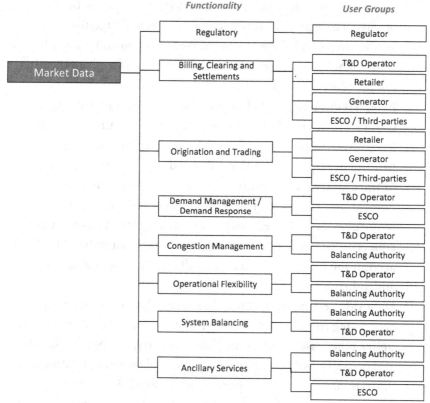

Figure 6.3 *(Continued)*

investments necessary for its collection, communication, and processing. Leaders must understand the four areas of data management to maximize use data for their companies:

- **Acceptable use and data access.** If the data is sourced or owned by another entity, leaders must be able to articulate why the data will be useful to their organization and how it will be used. Data management programs must be able to demonstrate the potential business impact they will make. Furthermore, access to the data must be available with manageable process constraints and technical constraints.
- **Changing requirements and emerging needs.** When data is used in combination with or in aggregation with other data sets, availability of all the data dependencies under a changing environment must be tracked. Often the lack of an end-to-end view results in concentrated

pockets of data and investments that don't deliver value because of gaps in data models, access, conditioning, and ingestion that are discovered only later, during implementation and testing. In addition, what works on a smaller scale or within a siloed function can fall apart during scaling.

- **External environment changes.** Data access, transfer, and storage are often subject to new rules and regulations that can have severe business implications. Recently, for example, a handful of large utilities that migrated data sets and analytics to third-party cloud-hosting premises had to reconfigure outsourcing efforts after new regulations and compliance requirements[3] enforced restrictions due to privacy and security concerns. Given that such changes are often unpredictable, accounting for possible changes will clarify usage modifications and valuation impacts.

- **Data value estimation.** While data is called an asset, standard accounting rules are not fully developed to value it as such. Data is unlike other assets such as the plant, property, equipment, patents, and even goodwill in that there is no formal valuation treatment laid down by accounting or corporate finance disciplines. Just because formal methods are lacking, however, does not mean that the value is missing. The onus is on the company to set up a consistent way to craft a business rationale or a business case for investing in data collection, conditioning, and storage. This process leaves much room for leaders to ascribe a valuation to data. But any valuation must consider the unique characteristics of data: that it can be shared without losing value, and indeed increase in value with more users and increased use; that its useful life is predicated on many factors that are not observable; that the valuable life of data is also use case dependent; and that if data is not available at the right time and situation, its value can be zero.

With these understandings in mind, leaders must ensure that their company's data management not only focuses on technology but also is tailored to the value created in the processes.

REUSE, REPURPOSE, AND STANDARDIZE

During the early days of advanced metering infrastructure (AMI), when most leading vendors used proprietary data structures and communication protocols, the executives at one leading US utility established a data management policy that the company would only accept data solutions using open protocols. It was a radical choice, given that the company had to forgo most leading AMI vendors and ended up selecting a startup company with very limited utility experience that used Internet Protocol, an open standard, for meter data communication. Within a few years, however, open standard became the dominant market practice that all vendors began providing. Leaders are not expected to get involved in the technical details of data management, but setting overarching guidelines that have strategic implications like standardization can provide clarity and direction, prevent waste, and expedite deployment. Such technical matters are often leadership issues for several reasons. When new technology is becoming available so rapidly, companies face a stream of new data and analytics products options promoted by vendors. Certain features and functionalities might be so alluring that an exception to standardization is made to suit a functional line.

But what if there are no formal standards available? In such a case, our advice is to find the de facto standards that most vendors and companies have adopted and to deploy them. Doing something unique in data structuring and analytics creates additional work when the data has to be shared and used by someone else.

At the core of data management are data semantics. *Data semantics* means grouping and using data according to an understood meaning of the data. Semantic data models provide a virtual representation of the data contained in physical systems that are exchanged between systems. Network service providers use these models for grid operations along with geospatial models, network models, device models, and communication topology. Models also map business processes and enterprise functions such as device management, contract management, and project management. The central ideas behind the creation of these models are that each category focuses on its own functionality, the interfaces are specified, and as long as everyone builds to the interface specifications, models can exchange and use data across models. The power industry over the years has agreed that semantic models are well suited for operational processes. Semantic modeling relies on a reference model that defines the entities and relationships, often referred to as the *ontology representation*, which is specific to the industry. The idea is that several entities and relationships are common across the industry. A transformer or a substation are entities, and their relationship in the power system is well established. By using the built-in entity properties and relationships, the utility can map the information in the physical world quickly and consistently to a virtual digital model. If the business wants to query the voltage and current of a transformer, the underlying details of the tags, connections, and even intermediary instrumentation devices like current and voltage transformers are hidden, while the functionality is already configured. Leaders do not have to become digital masters, but they must provide guidance on the extent to which their organization must use proven and standardized data to support business functions.

FOCUS ON BUSINESS OUTCOMES

The future business growth of the company will largely depend on how well the company is able to harness data and analytics to drive the top line and the bottom line of the business. As mentioned in Chapter 5, this capability largely depends on how well the company is able to serve customers in a personalized way; how it can improve reliability and safety without adding costs; and how it introduces new products and services to cater to changing customer needs. Investing in large data sets is of not much use if the data and applied analytics do not drive business value. This simple fact is prone to be forgotten when vendors of new systems demonstrate the vast amounts of data and the potential analytics that can be built. Moreover, it is only after a thorough assessment or a pilot that the dependencies on other data and systems are revealed. If parts of a data set are missing or not ready, or if the business processes are not configured, no business value is realized. The task of the business leader is to ensure that the data and analytics teams are working in harmony with the end goal of driving business value within a specific timeframe, and not with the goal of collecting and analyzing data for its own sake. Setting this guidance from the top ensures that criteria for commercial viability are applied and that a path to value is articulated when setting data collection and processing priorities.

Staying focused on business outcomes is not a new idea. Business-savvy companies often work backward from customer value to put in place everything that is needed. What is different here are the specific areas where leaders must be particularly concerned: vendor selection, new commercial models and value drivers.

Given the number of new technologies and vendors entering the market, vendor selection is a strategic issue. Gone are the days of a handful of global conglomerates. With new entrants and upstarts from outside the industry, and with interesting data and analytics offerings, due diligence is often complicated. A company having a nonstandard data structure and communication design may be attractive for a specific use case, but it may encounter configuration and adaptation issues if it becomes part of a larger system or a system of systems. Leaders must ensure that their teams are using an exhaustive criteria for vendor and system selection. Often, setting boundaries on the number of new applications or on the value thresholds helps with alignment among the business intent and vendor and system management teams. Such guidance can cover other related areas of strategic importance—such as data capacity planning, communication infrastructure, and application investments—that are necessary for cloud infrastructure design.

As the data ecosystem grows and multiparty solutions share and exchange data, business leaders must understand new commercial models and value drivers—often novel entities for legacy leaders—to respond to today's dynamic ecosystems that feature shifting roles of competition and collaboration, all affected by questions of business scale and growth, revenue and cost sharing, and rules of engagement. In many jurisdictions, utilities are finding that, more than technical challenges, commercial agreements are the real hurdles. One leading European company explored

Table 6.1 *Data Distribution Configurations*

Configuration	Characteristics	Pros	Cons
Bilateral agreements for decentralized contracts	• Electronic data interface-based message • Customized data processing • Unguided processes	• Customized to needs and interests • Limited number of parties in hub-to-hub communication • Less expensive to start • Easy to modify	• Extension and scalability challenges as Information and Communication Technology (ICT) interfaces grow exponentially • Dedicated resources • Manages customized interfaces for inter-hub interoperability • Deviation from de facto industry standards • Limited data access • Harder to monitor and oversee due to heterogenous agreements and lack of commonality • Asymmetrical bargaining
Consortium hub, a hybrid between decentralized and centralized	• Clearinghouse • Standardized messaging • Guided processing • Costs shared by customers	• Industry managed • Common and shared interest • Balances the pros of cons of a hybrid	• De facto control rights • Maintains consistency and transparency • Subject to greater scrutiny
Central agency, a fully centralized data exchange	• Centralized processing • Standardized messaging • Guided processes • Costs shared by grid operators	• Single, centralized hub • Minimal number of interfaces reduces operational complexities • Easy to implement industry standards and harmonization of business processes • Entry and exit • Shared pool resources for maintenance are overall efficient • Managed and controlled by a third party • Easier governance and regulatory oversight	• Neutrality • Expensive to set up the infrastructure • Third-party expenses • Resources for multiparty agreements • Longer learning curve for third party

the commercial basis for different data distribution models covering a range from a decentralized to a centralized structure. The tradeoffs were complex with long-term consequences, requiring leadership involvement, which was not apparent to the working teams in the beginning. Table 6.1 shows a snapshot of three different data distribution models—bilateral/ decentralized, consortium hub, and centralized—to illustrate this point.

Because commercial models rely on the availability of quality data, leaders must understand the implications of broader policies on fair use, privacy, and attribution. In the world of data and analytics, keeping the organization focused on business outcomes is one of the leader's main priorities. Leaders will have to draw on specific skills to manage data-driven business, particularly in negotiating contracts, forging partnerships, resolving issues related to sharing proprietary data, and creating business value.

OPTIMIZE FOR ROBUSTNESS AND RESILIENCY

Data breaches have already cost senior leaders—including many CEOs—their jobs. As the use of data expands, data sets from multiple disparate areas cross traditional functional boundaries, which increases the number of databases, data marts, sandboxes, and applications. Such increases in complexity bring exposure to new vulnerabilities. A single error can spread through multiple databases and create distortions with unknown consequences. Specific links and seams can be left unmonitored, leaving room for cyberattacks and security breaches. Tracing the data transfer paths and troubleshooting such errors can be expensive and difficult. Executives also often face the issue of *data drift*—unexpected and undocumented changes to data—in which the same data item (for example, the number of meter errors), yields different values from two different systems or departments. Data drift happens when data entry is not coordinated with data maintenance. Another problem that appears innocuous at the onset is the duplication of data. Every new system or application tends to pull data into its own database as a copy for its own application, creating another vulnerability that needs to be internally synchronized. Significant time and energy are spent on data conditioning rather than working with the data. Several central databases provide access to decentralized applications or one-off pilots and proof of concepts, which without coordination can leave vulnerabilities. But the increase in vulnerabilities and risks is no reason to not let data and analytics transform how the power industry operates. For sustainable use and enhancement of data and analytics, leaders must optimize on robustness and resilience. On the one hand, the interdepartmental silos, such as the walls between operational technology and information technology, need to be torn down, and on the other hand, there must be traceability to identify weak spots, and if there is a breach, quickly detect and take measures to limit the damage.

The topic of data reliability and resilience is not complete without a discussion about security. For every C-suite officer, from the CEO to CHRO and CTO, cybersecurity and, in particular, data security, is a major concern. When data is sourced from multiple systems, domains, functionalities, and organizations, ensuring security is critical. Security is ensured in two stages: first, during the design stage, and then, during the operations and maintenance stage. At the design stage, technical and administrative requirements must be met per legal and regulatory codes. Most security codes relate to confidentiality, integrity, availability, and possible removal'of data to ensure privacy for stakeholders and security of the operational, production, and delivery process including equipment, network, and systems. What leaders and specifically the CIO must ensure is that their organizations have a process in place that
(1) has the best available technology to identify a security event such as a data breach and (2) can quickly escalate and engage the right personnel, processes, and systems to isolate the threats, and then, once safe, can bring the organization back to its normal operation. The process must cover a broad range of stakeholder roles and responsibilities, must map who needs to be informed and/or consulted, and must identify who should make decisions on what issues and when. Data and analytics at scale will continue to unlock new opportunities for growth; to keep their companies relevant, leaders must embrace the new reality of ubiquitous data, multiple modes of data exchange, data complexity, and security guarding against cyberattacks.

FORMALIZE DATA GOVERNANCE

Data integrity is a basic expectation for data-driven decision-making. There must be no confusion regarding who is responsible for and who is the ultimate custodian of data integrity at various levels in the organization. Without data integrity, nothing related to data works. A robust and formalized data governance blueprint defines the rules and responsibilities of the parties involved in the handling of data, including the automated systems that control data generation and transfer such as data sources, network, storage, conditioning, processing, update, and destruction. Stakeholders responsible for the various activities related to data handling should be identified and their roles defined. Data governance lays out the ground rules for (1) conflict resolution, (2) lines of hierarchy and authority, (3) allocation of tariffs and costs in multiparty situations, and (4) the processes and rules for change management. All of these activities may appear routine and blindingly obvious, but it is only natural to avoid adding overheads and to wait until a breach occurs to react and make these changes. In the US, the North American Electric Reliability Corporation Critical Infrastructure Protection standards are an example of one of the early attempts to provide a framework for cybersecurity and governance rules for data related to critical systems. Several AMI programs in the US faced legal challenges on customer privacy, causing program delays, jamming management time, and resulting in expensive litigation proceedings. Data collection, data exchange, data analyses, and fair use of data have created an ongoing debate regarding access, availability, contracts, and monetization of data. Worldwide, the regulatory and legal frameworks are constantly being updated to meet the challenges that emerge as different grid modernization programs are implemented. The emergence of European Union's General Data Protection Regulation is one such example.

Data governance is closely linked to innovation and the future positioning of a company. Thus, the company will need to set guidelines based on a range of strategic factors: business priorities, risk management, transparency, security, and trustworthiness. Rules need to assure data integrity and must balance technical and business performance. Too

Table 6.2 Examples of Specific Roles and Decision Rights for Data Management Data

Process	Governance Activities	Sponsor	Governance Committee	Business Analyst	Data Technical Analyst	Data Owner	Architecture Committee	IT	User	Source
Leadership	Funding and sponsorship		A, R	I	I	C	I	I	I	I
	Prioritization of activities		A, R	I	I	C	I	I		
	Development of initiatives	I	R	C	C	A		I		
	Performance measures	I	I	A	A	R		I		
Program Management	Definition of functional teams	A, R	A, R	I	I	C	I	I		
	Identify data steward(s)	A	C	I	I	R	I	I		
	Performance management		A, R	C	C	C	C	C	C	C
Data Management	Data retention		I	A	A	I		R		
	Lifecycle management		I	A	A	C		R		
	Business rules alignment		I	R	R	A		C		
Processes	Issue resolution	I	R	C	C	A	C	I	I	I
	Change management	I	R	C	C	A	C	I	I	I
	Conflict resolution		A	R	R	C		C		

R—Responsible: Person/group/committee has responsibility for direct delivery of the task/deliverable.

A—Accountable: Person/group/committee holds the ultimate responsibility for the completion of the task/deliverables.

C—Consult: Person/group/committee that is consulted for a task or deliverable for guidance and alignment.

I—Inform: Person/group/committee that needs to be informed of the task or deliverable.

many controls and restrictions can give rise to unnecessary bureaucracy and affect efficiency. Too much openness can create risks regarding intellectual property. Given that data must be bound by rules and access rights, leaders can easily be overwhelmed by the multiple factors to be considered. To cut through the complexity, it is helpful to categorize governance issues that directly pertain to strategic objectives and to establish principles for the rest of the organization in areas that are more functional and operational in nature. Many companies create a data governance committee to handle cross-functional matters; specific roles and decision rights are carved out for committee members. Leaders must steer the way in creating such governance bodies and vesting powers in them to make decisions based on established criteria and facts that are forward leaning. Table 6.2 provides an illustration of definition of roles and responsibilities related to data governance

Under VUCA conditions, governance models need to be adaptive to be able to respond to emerging technology and data requirements. Electric vehicle tariffs are a good example of the evolving nature of technology and commercial structuring that vary even at the local level. Data governance cascades into data administration procedures for commands and settings, tuning rules-engines in network models, device management and configuration, scheduling, version control storage and archiving. High-level governance policies need to align with functional-level procedures; a weak link anywhere in the system can put all entities in the system at risk. It is the task of leaders to ensure data robustness and resilience, and all in the company must know that data governance is everyone's business: it is not limited to a particular group but is an enterprise-wide concern required for companies to succeed in the emerging power industry.

SUMMARY

L eaders must provide policy support and strategic direction for their company's data management; without proper data management, a utility will not be able to embark on its necessary modernization journey. Leveraging data for decision-making and optimizing operations should be a leader's top priority, and thus leaders must ensure that data management is framed in terms of business value. To do so, the focus for the organization is on driving clarity regarding data sources, ensuring data access and availability, and estimating value impacts. Often this focus leads to friction and resistance from those who are responsible for technical and functional details and may not always see or understand the value of data in the overall company architecture. Threats to cybersecurity and related vulnerabilities create additional risks that data management must address. Undesired parties having access to data can be threatening to an enterprise. By setting expectations on the use of standard over one-of-a-kind solutions, leaders can keep investments targeted to business objectives and can reduce wastage of resources. Under uncertainty, keeping the organization tightly aligned to the target also means optimizing for resilience and robustness. Leaders have a large and direct role to play in setting the culture for the entire organization in how to manage data; in doing so, leaders must champion engagement with various internal and external stakeholders, establish data governance, set the rules of engagement, and seek regular updates to ensure the organization is not just collecting and storing data but is driving value from it.

NOTES

1 "The World's Most Valuable Resource is No Longer Oil, but Data," *Economist*, May 6, 2017, https://www.economist.com/leaders/2017/05/06/the-worlds-most-valuable-resource-is-no-longer-oil-but-data. Clive Humbly, *Association of National Advertisers Summit* (Chicago, IL: Kellogg School of Management at Northwestern University, 2006).
2 The most common example is the Open System Interconnection seven-layer network model.
3 Sam Higgins, "Return of the Sovereign Cloud," *Forrester*, July 29, 2020, https://go.forrester.com/blogs/return-of-the-sovereign-cloud/.

PART THREE

Preparing for Execution Readiness

CHAPTER SEVEN

Rethinking the Business Case

What Are the Requirements to Develop Business Cases Differently?

DOI: 10.4324/9781003353997-11

The business case is the linchpin document that justifies the costs incurred to meet mandated business performance targets: reliability, affordability, and shareholder value. Developing business cases is a routine practice for regulators, boards, and managers when making investment decisions. So, it is worth addressing upfront why dedicate a complete chapter on this topic. Although there is nothing inherently new in the practice of business cases, under VUCA conditions, the traditional approach to business cases often does not work that well. Current investment portfolios consist of programs that feature emerging technology, many unknowns, and limited historical information making business case development as much art as science. If a business leader is asked where to allocate a marginal hypothetical dollar, it is not easy to decide whether to spend it on operational productivity or on a new venture. In many situations, companies have not adopted a robust and consistent approach to building business cases, and the result has been catastrophic.

While the financial measures of internal rate of return, net present value (NPV), and payback time will remain the metrics through which investment decisions are made, the inherent uncertainty in the business environment raises questions on underlying assumptions—specifically on of what postulates must be true, if the projected results are to be believed. That said, emerging technology and uncertain conditions are not unique to the power industry; most businesses face these challenges. In response, leaders are on one extreme spending disproportionate efforts and resources on detailing costs and benefits estimates despite the fact that today's dynamic conditions almost by definition impose limits to data accuracy, which in turn drives large variances between planned and actuals. On the other extreme, leaders are making investment decisions based on gut feelings without a firm grounding on numerical estimation, resulting in their companies taking enormous risks on return on investment. In this chapter, we focus on creating business cases under uncertainty, and we provide a framework that rests on a data-driven business narrative that provides a rational basis for future unknowns. We rely on the basic principles of corporate finance[1] and economics in this discussion, but this chapter is not about the mechanics of program valuation or benefits-costs analysis but rather on the art of crafting business cases in the modernization context.

EMERGING ISSUES WITH BUSINESS CASES

Today, most business cases are designed to value projects that are incremental developments with largely predictable outcomes. The projects follow well-defined specifications that do not vary much from planning to execution. But imagine a situation where government policy mandates a utility to move from fossil fuels to cleaner resources within a period of twenty years. The utility employs multiple strategies—a key one being increasing the number of renewables-based distributed energy resources near the points of consumption. The outcome of this strategy is highly uncertain because the DERs are owned by different third-party aggregators and customers, meaning achieving the utility's objectives depends on forging partnerships; facilitating procurement, installation, and interconnection processes; and creating efficient commercial options. Investments include a variety of technologies such as DER management systems, load-balancing mechanisms, grid upgrades with self-healing protection and control systems, advanced voltage control, customer portals, various customer communication channels, maintenance and support services, and billing and fulfillment services. Many of these technologies are constantly evolving, sometimes changing rapidly in ways that are not always predictable. Many of the vendors and suppliers are upstarts with no installed base or survival guarantee to withstand business cycles.

There are many ways business cases can go wildly off target. In recent years, we have worked with business cases on technologies that did not match with reality at all, leading to making wrong bets. For instance, hardly anyone predicted the 90 percent cost reduction in solar PV and battery costs over a decade and the subsequent growth of solar and distributed energy resources. Furthermore, number projections based on accounting can serve as a reasonable basis to determine benefits and costs. But an approach to projecting future costs, anchored on accounting data, runs into several limitations with the modernization projects under VUCA conditions. Similarly, monetary benefits derived from outright savings, productivity improvements, and increased revenue from current baseline values can be misleading if limited by the lack of forward-looking insights. In such conditions, Business cases need to consider the following:

1. *Modeling future project economics* (growth, expected cash flow, and risks) that are highly uncertain but are required to match the "reasonableness" and "prudence" criteria necessary in a regulated business.
2. *Adjusting for affordability limits* that constrain the size and pace of investments given the usual front-loading of costs and a lag in benefits realizations.
3. *Anticipating frequent and disruptive changes* in digital solutions such as platforms, data analytics, software-as-a-service with short system lifecycles, and anticipating economics driven by third parties outside the power industry.
4. *Incorporating large numbers of first-of-a-kind projects* that offer no prior history of value, costs, and benefits.
5. *Attributing benefits and costs of localized projects* (DERs, microgrids, and self-healing circuits) to ensure that the costs are allocated fairly to beneficiaries.
6. *Anticipating the entry of new and nontraditional players* with innovative offerings, forcing frequent changes in business plan assumptions.
7. *Managing biases toward projects* that may be easier for immediate regulatory approval and rate base growth but not optimal for the overall vision of the company.
8. *Balancing short-term shareholder returns* (quarterly/annual earnings per share [eps] growth) with long-term valuation and risks of the company.

9. *Reconciling enterprise value with economic value* created but not captured, such as benefits captured by customers and third parties ("societal benefits").

10. *Accounting for strategic options* created from the investments that are not quantified, such as value of experimentation, market readiness, and options to stay on course-to abandon the project, or to mothball the project based on learning effects during the course of the project.

AVOIDING TWO DEADLY TRAPS

The complexities of these conditions can be daunting and even unsettling. As a consequence, business cases often fall into two common traps—"do nothing" or "drown in detail"—in their quest to gather and seek greater accuracy and precision. Both approaches can hamstring modernization programs and can even have disastrous consequences:

1. **"Do nothing."** When companies working on business cases are confronted with value estimates that are highly speculative, costs that are uncertain, unavailable historical data, the knowledge that past events do not provide a basis for the future, and a lack of analogs and comparables, it is not uncommon for them to stall. The management of one utility, that faced these conditions while developing a business case for customer experience, decided to "do nothing" until more conclusive evidence for the case could be gathered. The utility did not analyze the "do nothing" scenario because the usual practice to estimate monetary benefits on an avoided cost basis was not easy. The range for total costs for deployment was too wide and was dependent on customer adoption, which was speculative. Within a few months of management shelving the initiative, a major outage took place. The utility was slammed with customer calls, which the traditional interactive voice response (IVR) systems and customer support processes were not

designed to handle. These practices were out of sync with customer expectations, and irate customers, whose basic service queries were not answered, took to social media. The stock price for the utility fell, followed by multiple regulatory inquiries and punitive procedures.

Most of these problems could have been avoided if the utility management had analyzed the "do nothing" scenario—not just from a monetary perspective but also from a qualitative risk perspective (in economic terms, analyzing the "do nothing" case for opportunity cost of not doing the project). Will the utility in question lose customers, lose money, incur damage to its image, or experience a negative impact on its business? For example, will customers move to another provider if a wholesale and retail choice exists? If a commercial customer wants a new connection, what is the likelihood that a neighboring utility's offering will be more attractive? As customers are increasingly valuing attributes beyond the price of electricity, such as reliability, resilience, quality of customer service, and range of customer choices, business cases must consider additional factors such as value preservation, sunk costs, and real options.[2]

2. **"Drown in detail."** In the face of uncertainty and ambiguity, leaders often pursue the approach of relentlessly drilling into details in the hope that they might discover a "silver bullet" that makes their choice clear. Many CFOs and leaders keep asking for more granular information and evidence, resulting in irrelevant analyses, churn, and confusion. Often the quest to get data from who has done similar program before can lead to a witch hunt, when the deployment is first-of-a-kind or site-specific. In recent years, as the scrutiny of grid modernization investments has grown, regulators and management have often fallen into the trap of seeking greater granularity and details. Under uncertainty and ambiguity, increased details and granularity can obfuscate what is important and create distortions, drowning teams and decision-makers in the trees, where they risk losing sight of the forest. Despite providing a (false) sense of precision, such studies do not provide insights into future valuation.

To overcome this trap, programs are better off starting with a realistic assessment of the accuracy that is possible with reasonable effort for all the components of the business case: cost, schedule assumptions, savings estimates, productivity gains, and risks. The business cases can term this uncertainty in model accuracy as "model risk." Programs can then use thresholds on the model risk to make decisions. Where the risks are out-of-bounds, instead of drilling down into smaller parts, programs need to find other ways (such as applying reasoned judgments and beliefs; or employing a systemic method of controlled experiments, proof of concepts, active observation, and sense-monitoring the landscape) to deal with risk assumptions and uncertainty. Leaders must accept that uncertainty introduces modeling risks that cannot be swept away by drilling into noisy data that is not tightly correlated to the outcome. Business cases must be constructed as transparently and as clearly as possible so that risks are made evidently clear and transparent for the decision-makers. Beyond that, leaders need to make decisions based on their own collective wisdom and good judgment.

DATA-DRIVEN BUSINESS NARRATIVES

Current industry conditions call for a modified approach to business case development. Business cases have to be built on a believable narrative of the future that is supported logically with the best available information and with numbers backed by business logic. Developing a business case is not an exact science, and value is not fully derived with accounting precision. Business cases traditionally separate quantitative from qualitative benefits, with quantitative benefits mainly focused on the company's financials, such as earnings and enterprise value. Quantifiable benefits drive most investment decisions because the focus is usually on earnings per share and enterprise value. Qualitative benefits, which do not offer a direct line to net income and returns, are included but rarely override quantifiable benefits in decision-making, other than in extremely specific risk situations—for example, cybersecurity and cascading grid failures. Historically, this bias toward financial certainty has caused power companies to fill their investment pipeline and investment portfolio with projects that have certain outcomes (e.g., positive NPVs), meaning a discouragement and de-prioritization of first-of-a-kind projects and risky projects. But power companies today require different kinds of projects to meet future performance goals, demands, and pledges; they increasingly include projects on sustainability, digitalization, climate change, and renewables—projects that are weighted more toward qualitative and strategic benefits than immediate financial returns. To handle these hard-to-quantify investments, power company business cases need to combine qualitative and quantitative components to create a narrative that (1) is well-constructed, cohesive, and believable; (2) includes a realistic pathways-to-benefits realization; (3) is data-driven and evidence-based, with a quantified representation of reality; and (4) provides a flexible model that is adaptable for possible cost and benefits changes.

Depending on the internal practices of the company and the regulations that apply to it, business case justifications today fall into two extremes: (1) relying mostly on the narrative to prove value or (2) relying on numbers to prove value. The narrative-focused business case depends on storytelling. These campaigns employ a consistent, repeated, universal, inclusive, and inspiring message. Numbers-focused business cases, by contrast, hinge on the estimated quantitative values of investment amounts, expected benefits, and returns in terms of savings and revenue, which aggregate to growth in earnings, dividends, and enterprise value. In one large modernization program, where the goal was to create a sustainable smart city, the business case was fully reliant on the narrative of creating a modern grid. The stated envisioned future for the city overrode the need to go through a rigorous process of quantifying the costs and the benefits to guide decision-making. Following heavy initial investments, when the first set of financial benefits were realized, the large shortfall between planned and actual benefits took the leadership by surprise. Management abruptly applied brakes to new investments, resulting in good projects being abandoned and the overall program underachieving. In contrast, focusing too much on numbers, as seasoned leaders know, can also harm programs. As stated earlier, strategic projects with future-oriented costs and benefits are particularly hard to quantify, which can either disincentivize managers from pursuing such projects or can tempt them to dress up assumptions for more optimistic results. Good business cases require a combination of narratives that support the plethora of assumptions and define a pathway to benefit realization, consistent with the value realization narrative for costs and benefits.

Another potential pitfall is when business cases treat a program as a discrete set of initiatives or projects to be delivered in project teams working in silos. Unless this approach is well-coordinated, it can result in several issues: it can miss both positive and negative synergies among projects, it can overlook dependencies between systems and processes, and it can lead to poor sequencing and timing of investments. Often these issues also result in programs double- and triple-counting benefits and costs or not accounting for integration efforts. In several companies, we found the same pool of workforce reduction getting credited in AMI, distribution automation, and mobile workforce management. By using a narrative, business cases can bind all the different components into a consistent description of the portfolio—how they interact at the individual project level as well as at the overall program level. The narrative approach provides clarity for decision-makers and leaders, and it helps to maintain the connection between the forest and the trees. In addition, a data-driven narrative allows for the business case to serve as a living document that can be reviewed and altered when situations change, and when what was once believed to be true no longer holds. It makes it easier to test, discuss, debate, and adjust to suit the changing environment.

A business narrative that is transparent about its assumptions and is supported by numbers that make sense is easier to communicate and has a better chance of influencing investors and regulators. For example, investors rewarded Amazon and Tesla for risk-taking and innovation, for years, way more than they did for legacy companies because their business growth narrative was simple, logical, and credible. In a changing environment, if the underlying assumptions change, and are no longer relevant, then the business narrative must also change. While it may not be obvious, many organizations that approach economic and financial estimates with a traditional, predictive mindset find this truism uncomfortable, and hard to practice with discipline and diligence. Rather than adapting the approach to VUCA conditions, many companies try to justify and hold onto the original estimates, take punitive actions against the estimators, or distort facts all exposing the program to disastrous outcomes.

FRAMEWORK AND TESTS FOR A BUSINESS CASE

With these considerations in mind, we present a framework for developing a business case that features a business narrative suited for valuation under uncertainty and limited information (Figure 7.1). Developing a business narrative and then translating to numbers is not a novel concept, and the framework described here is adapted and drawn from the ideas discussed in detail in Damodaran[3] and Denning.[4]

The three main steps involved in creating a business case are iterative.

1. **Develop the business case narrative.** The narrative articulates the outcomes that the business case wants to model, the deployment speed and timing of specific projects, the reasons the projects will work, and the uncertainties/risks that may impact the outcome. Writing a business narrative can be complex; by far the most important aspect of it is the valuation exercise. Complexity stems from the fact that there are multiple inputs, data sources, and stakeholders that are intertwined and inextricably linked. The narrative must be cohesive and internally consistent, which requires identifying and reconciling all conflicting points, including explaining why a certain course was chosen over others. Business cases must remove bias and rely on facts and data as far as possible. Narratives can easily be swayed by moods, the burning issues of the

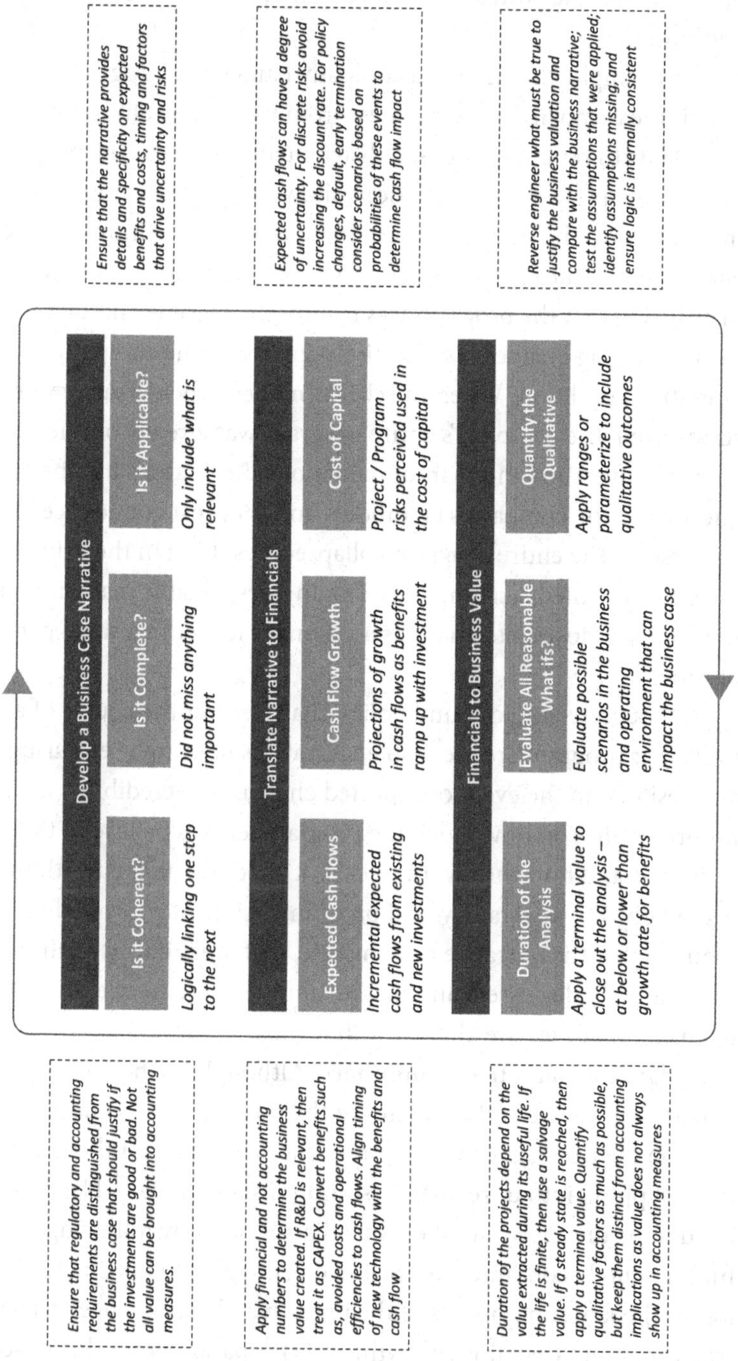

Ensure that the narrative provides details and specificity on expected benefits and costs, timing and factors that drive uncertainty and risks

Expected cash flows can have a degree of uncertainty. For discrete risks avoid increasing the discount rate. For policy changes, default, early termination consider scenarios based on probabilities of these events to determine cash flow impact

Reverse engineer what must be true to justify the business valuation and compare with the business narrative; test the assumptions that were applied; identify assumptions missing; and ensure logic is internally consistent

Develop a Business Case Narrative

Is it Coherent?	Is it Complete?	Is it Applicable?
Logically linking one step to the next	Did not miss anything important	Only include what is relevant

Translate Narrative to Financials

Expected Cash Flows	Cash Flow Growth	Cost of Capital
Incremental expected cash flows from existing and new investments	Projections of growth in cash flows as benefits ramp up with investment	Project / Program risks perceived used in the cost of capital

Financials to Business Value

Duration of the Analysis	Evaluate All Reasonable What ifs?	Quantify the Qualitative
Apply a terminal value to close out the analysis – at below or lower than growth rate for benefits	Evaluate possible scenarios in the business and operating environment that can impact the business case	Apply ranges or parameterize to include qualitative outcomes

Ensure that regulatory and accounting requirements are distinguished from the business case that should justify if the investments are good or bad. Not all value can be brought into accounting measures.

Apply financial and not accounting numbers to determine the business value created. If R&D is relevant, then treat it as CAPEX. Convert benefits such as, avoided costs and operational efficiencies to cash flows. Align timing of new technology with the benefits and cash flow

Duration of the projects depend on the value extracted during its useful life. If the life is finite, then use a salvage value. If a steady state is reached, then apply a terminal value. Quantify qualitative factors as much as possible, but keep them distinct from accounting implications as value does not always show up in accounting measures

Figure 7.1 Business Case Framework

moment, and they can slip into a zone that ceases to be credible or relevant.

For example, the CEO of a company launched a program to switch the company's core business from large, centralized fossil fuel plants to renewables-based distributed resources. The program became the perfect public relations showcase for the company's march toward the future. Because there was overwhelming public endorsement, the company was reluctant to expend much rigor in ensuring that the program was technically feasible and that the underlying business assumptions applied to the company's operating condition. When the claims made in the narrative were scrutinized, the company's entire program was already off the rails with cost overruns and margin reductions. Regulators and the public questioned the company's credibility, and they lost confidence in leadership. The entire program collapsed, resulting in the firing of the CEO, the media calling the CEO an irresponsible maverick, and the company losing its image as a visionary leader in a short period of time.

Although technology and policy changes over the course of a multiyear program are inevitable, narratives need to be enduring and resilient in the event of expected change. The credibility of a program relies on how much the program leaders believe in their narrative, and how they work to keep it alive and aligned with their overall objectives, strategy, and program vision. Staying credible requires that the narrative be complete, that areas of uncertainty are transparently discussed, and that assumptions are frequently tested to ensure that they are aligned with reality.

2. **Translate the narrative to numbers.** Although the business case narrative gives life and character to the business case, numbers provide the necessary measures and checks to prevent the narrative from slipping into unrealistic assumptions and fantasy. Narratives need to be driven by numbers in a disciplined way following fundamental economic and financial principles and business logic. The key components of the narrative ultimately need to be translated into cash flow, growth, and risk parameters. This process

is not an exact science; by design, several qualitative aspects are translated into quantitative constructs using business logic. Many quantitatively oriented people find it uncomfortable to make such a leap and seek to alleviate their discomfort by looking for more details and precision as discussed earlier. Seasoned practitioners in financial valuation are experienced in these situations and know a judgment call is necessary for the narrative-to-numbers translation. The important point is to focus on the rationale that links the story to the numbers, so that, as new information emerges, the story can be changed along with the numbers. Business cases must constantly test by asking the question of what has to be true for a belief to hold. If the business risks are perceived to be higher than before, then the discount rate used to run a discounted cash flow valuation will be higher than the usual cost of the capital for the company's core business. The ramp-up rate of the program and even assumptions regarding qualitative expectations (i.e., how many customers will adopt new initiatives or services) can be brought into the growth rate number as well. On the cost side, similar estimates can be made to convert the amount of time spent on traditional activities such as marketing, service calls, and design and engineering into saved time, improved productivity, and freed-up resources by deploying a more advanced solution. This entire process may appear messy with multiple opinions and discussion, but that is the nature of this step, and it works so long as company leaders and regulator (1) share a common grounding and understanding of where the numbers originate and (2) are in general agreement with the underlying assumptions.

3. **Financials to business value.** Once the financials are developed, running the business valuation exercise can rely on standard methods such as discounted cash flow with the appropriate terminal value for the programs. It is important to have an exit point for the analysis. Often, not much thought is given to this question, an oversight that can seriously impact the business case either when the project or the program keeps going into perpetuity or when it finishes its lifetime. Depending on the situation, valuation

models need to assume a salvage value and a terminal value. These assumptions are particularly important for technology-based projects within a modernization program. The maturity cycle for projects or technologies is sometimes under five years, and in ten years the entire generation is out of date, out of support, and out of market—a situation that contrasts sharply with those of electromechanical legacy assets, which run for forty to sixty years and have a different valuation profile.

Business case developers engaged in valuation exercises need to keep testing the consistency and robustness of their logic under various real-world changes in business and operating conditions. One effective way to test valuation estimates is to reverse-engineer from the valuation. For example, if the business case assumes a growth rate of 3 percent, asking what assumptions must be true in the business narrative to support a 5 percent growth rate will prompt an inquiry and/or debate, and will ultimately validate the valuation. To achieve 5 percent growth, do more customers need to sign up? Do electricity rates need to go up? Do more customers need to find benefits in the demand response program? Or must a new policy be introduced? An exhaustive take on these questions to test for consistency and congruence of the business narrative will uncover implied assumptions and expose hidden causal links. Any inconsistencies found must be addressed. The more rigorous this exercise is, the better the program understands the risks and the components that matter to drive value.

With such exercises, we have found that the choice of discount rate is important in the context of risks and risk premium. It is not unusual to debate over it and often the uncertainty of cash flows is captured in it, which is not a good financial valuation practice. In our experience, cash flow has a greater sensitivity to valuation in most situations and often varies widely from the estimates. Delays in benefits realization, avoidance of accounting process changes, and sensitivities of technology performance impact the expected cash flow and sway the business valuation more than the discount rate. Therefore, when it comes to testing the valuation parameters, understanding the cash flow over the project lifetime should be prioritized over discount rates.

CASE STUDY: APPLYING THE BUSINESS CASE FRAMEWORK

A large utility that operates as a government entity is considering a huge investment of billions of dollars to modernize itself as a smart 21st-century company. The goal is to position the utility as the leading sustainable utility in the world by employing the best available technology. The investment has a strong image and public branding motivation, but the investments also have to be prudent, efficient, and reasonable for customers, who will have to bear costs through higher electricity bills. Hence, the investments must generate benefits that exceed costs. What are the possible business case approaches? The utility faces three possible scenarios as it charts its course:

1. **All-in big bang deployment to 100 percent clean energy.** In this scenario, the utility embarks on a road to accomplish 100 percent clean energy transition by a certain date. Interim milestones are set up leading to the end goal. Funding is not a constraint because the source of funds is assured so long as the interim milestones leading up to the end goal are achieved and the funds are spent prudently, meaning good management and governance ensuring that the program delivers on its milestones. The last 10–20 percent of emissions reductions are not expected to be met with current technologies; therefore, appropriate R&D investments must be made to ensure solutions are ready.

2. **Affordability based.** In this scenario, the utility has an affordability limit that puts a ceiling on how many investments can be made to replace and replenish assets currently in service and to recover costs through annual increases of customer rates. Capital allocation is one of the central functions of the C-suite, as is the ability to

fund compliance, reliability, and safety projects along with a robust pipeline of modernization projects. After "must-do" investments are made, budgets are allocated to the program on a yearly basis to support modernization.

3. **Level of change.** In this scenario, the utility has a limited capacity to absorb process and technology changes. Because the effort and time required to make changes in legacy infrastructure and processes can be substantial, including the adoption of changes, only limited top-down authority can be exercised. Personnel are allowed time to adjust, train, and adapt. Asset investments are in line with the regular turnover of existing assets.

The business drivers for the program include:

- **Rate base growth:** Adding new investments for continuous rate base growth to increase returns. These investments include asset upgrades, DER infrastructure, grid automation, real-time sensing, monitoring and control, flexible AC transmission systems, and other capital investments.
- **Value-added services:** Adding new revenue streams from electric vehicle infrastructure, online marketplaces, connected home services, and enhanced data services.
- **Commodity sales:** Moving from electrification of nonelectrical sectors such as (1) electric vehicles, (2) electric heating and cooling, and (3) indoor farming.
- **Operating margin (pre-tax operating income):** Realizing benefits from various special and continuous improvement programs that increase programs productivity and from process efficiency that reduces operational expenses.
- **Capital efficiency (return on invested capital):** Managing assets that create returns from optimal maintenance and operations.
- **Reliability:** Making improvements in the System Average Interruption Duration Index and the System Average Interruption Frequency Index scores because of operational improvements.
- **Safety:** Avoiding major safety events including fatalities and Occupational Safety and Health Administration recordables.

TRANSLATING THE BUSINESS CASE NARRATIVE TO THE PROGRAM DESIGN

Once the business case narrative has been established, it is time to translate the narrative into a design for the program. In the example in Table 7.1, the story described in the left column is translated into a program design in the right column.

TRANSLATING THE PROGRAM CONSIDERATIONS TO NUMBERS

When designing the program, a series of underlying conditions have to be met—or believed to be true—to estimate the costs and the benefits for the business case. In some cases, the numbers are available, and in other

Table 7.1 *Narrative to Program Design*

Narrative	Program Design
The utility is core to the government's sustainability agenda, which requires reducing greenhouse gas emissions using the best available technology as quickly as possible.	Deployment schedule for demand-side management, energy efficiency, renewables growth, fossil retirements, and customer satisfaction is critical and planned for minimum time to achieve commercial operation.
The utility has no major resource constraints; however, the investments have to be prudent and backed by a reasonable business case.	Assumes funding is unconstrained at the current debt-to-equity ratio. Rationale for investments must be justified with benefits outweighing costs.
Risks of nonperformance are elevated due to high visibility and the importance of the programs.	Offers robust technology selection, validation, and execution process.
Automation and use of the latest technology need to be advanced not just for benefits but for overall branding around the world's leading digital utility.	Where possible, applies advanced technology that meets the threshold requirements for performance and risks.
Customer satisfaction including the front-end interface with the utility is expected to be best-in-class using all possible digital and modern design technology.	Customer programs are designed to be on par with leading customer experience providers outside the utility industry sector—e.g., banking, retail, and IT.

Table 7.2 *Example of Valuation Inputs per Assumption Area*

Assumption Area	Consideration	Input Basis
Program deployment, timing, and ramp-up	Aggressive deployment and ramp-up	Industry averages and how do they compare
Cost assumptions	Cost premium paid to obtain the best available technology	Costs from vendor data/quotes to reflect first mover, special cost-sharing programs, and implementation and integration of cost uncertainty
Benefits assumptions	More weight is placed on brand and image building, and on customer satisfaction	Knowledge, awareness, relationships, and experience ease doing business—time-savings, effort, first-time resolution, integration; adjust benefits to cost ratio targets
Risk assumptions	Execution and performance risks to manage complexity	If a higher percentage of failure to deliver the benefits is expected and is reflected in a cash flow adjustment and/or a higher discount rate

cases, they are not and must be assumed. This process can be complex because each input has to be created based on assumptions. Table 7.2 presents an example of such valuation inputs.

These valuation inputs can be used to create a detailed view for a business case, including, for example, the estimated implementation timeline; the variation in hardware, software, service, and O&M costs; the one-time labor savings; the recurring labor savings; the nonlabor savings; and the expected revenue increase.

To conclude, it is worth remembering that models used for business cases are usually economic valuation models and are different from the models that are used in natural sciences and engineering. Natural sciences and engineering models are used to make predictions, but a good model for a business case is not a predictor of future events. It is at best a construct to provide insights and guide thinking by summarizing the different cost and benefit drivers at play. A cost-benefit model encapsulates the collective intuition of the operating context—the regulatory and the market environment, and the internal operational arrangements. These intuitions are often qualitative, which are then quantified. When economic valuation models are run, they shed light on which phenomena or drivers matter, leading to further exploration. Benefits-costs models when done right provide insights on ramping, waiting, benefits realization timings, project returns, cash flow, and more. Leaders must keep in mind that these models are not created for making definitive real-life predictions. The information these models provide about the future—e.g., future benefits, investment returns, payback period—must not serve as a crystal ball to foretell the future. Models are useful because they provide an excellent way to understand and communicate the complex relationships among the value drivers, resources, and outputs, which supports in a wide range of management decision-making.

BUSINESS CASES FOR UNREGULATED INVESTMENTS

In unregulated markets, business cases can vary greatly depending on the capital structure and financing of projects. Projects can be financed by balance sheets, or financed using nonrecourse debt. Unregulated merchant generation mostly backed by project finance took off in the 1990s and 2000s in the US, Australia, and the UK and has gone through ebbs and flows. In recent years, projects have developed innovative financial structures to support the financing and growth of renewables. Renewable energy financing is a highly complex area that exploits innovations in accounting, tax laws, and project finance to accommodate the different financial and tax incentives provided by policymakers across the globe. Finance professionals are familiar with these structures, but given that a full treatment is outside the scope of this book, we limit our discussion to an overview of a few of the basic concepts.

First, the primary difference between a regulated utility-based investment in traditional fossil-based generation versus a renewable DER, either solar or wind, is that utilities treat revenues (volume multiplied by price) of regulated generation to be "as-generated," which means they themselves absorb all the risks of variable production. For private developers, who lack a strong balance sheet, lenders limit their credit risk contractually by ensuring enough revenue and cash are available for debt service, and offtake is contractually guaranteed by power purchase agreements (PPAs). Construction costs are fixed using lump-sum-turnkey contracts with the engineering, procurement and construction, and often O&M is fixed using service agreements.

Second, to take advantage of policy incentives, the financing structure of renewables becomes contractually complex. In Germany, there were feed-in tariffs that provided the economic impetus to build and adopt cleaner resources. In the US, the production tax credit and investment tax credit, along with accelerated depreciation, provided tax benefits. Because many sponsors and project developers did not have the taxable income to take advantage of the tax benefits, the industry used innovative financing structures by forming private partnerships (special-purpose vehicles) to bring in tax equity in return for tax benefits. Because income and cash are not the same, the deals are structured such that the tax equity partner, which has enough income to use the tax incentives, drew most of the income and used the tax benefits in the early part of the project life cycle when the MACRS[5] and tax incentives are dominant. Once the target return set contractually for the tax equity investor was achieved, which may be after ten to twelve years, its equity would either be bought out by the other equity investors, or its stake decreased considerably. The cash equity partner remained as the only equity holder for the duration of the project. As this example shows, the interplay of tax equity, cash equity, and lenders makes renewable financing complex.

Third, as renewables' penetration increases, covariance risks increase, which increases related risk hedging complexities. Covariance risk is the relationship between generation and price. For instance, as experienced in Texas in February 2021, where the penetration of wind generation was high, a systemwide weather event shut down enough wind generation to cause real-time market prices to increase and hit the cap of $9,000/MWh. In such a case, a particular renewable resource is unable to take advantage of the higher prices because of the common failure modes such as weather events. At the same time, it has to buy the shortfall in a spot market at exorbitantly high prices if it has a firm contract with its offtakers. Given the nature of the risks, it turns out that additional costs of insurance are needed to limit financial exposure. As capital markets, technology, and policies evolve, financial institutions are creating innovative financial instruments such as synthetic or virtual PPAs, balance of hedge,[6] lines of credit, and contracts for differences. Fourth, the shape of the demand profile adds to the risk impacts on the business case. The shape of the demand profile of an intermittent supply from wind and solar introduces what is called the *shape risk*. A data center may enter into a corporate PPA contract with a wind supplier on an "as-generated" basis, which means the data center will be consuming what the supplier is producing. However, if the generation cannot match the demand profile (what is known as *change in load with time*), then the data center has to fulfill its requirements from other sources to ensure continuity of its operations. These events can strike a blow to the business case, and therefore these programs must implement limits on its financial exposure.

Fifth, financial contracts introduce *basis* risks: For renewables, synthetic or virtual PPAs are common structures that convert floating payments to fixed payments based on a strike rate. A renewable generation project may be selling to the merchant market at spot rates. To hedge the revenue (price times production) uncertainty, the project executes a contract with a financial institution to swap the spot rates for a fixed rate at a certain strike price—similar to swapping a floating interest rate for a thirty-year fixed mortgage for a house. With electricity, however, the node at which the physical power is sold is different from the reference location of the contractual strike price, which is anchored to a hub. This structure introduces a risk, and projects must absorb that risk because it cannot be hedged away. Such exposure can affect the financial valuation of the project. All of these elements of the unregulated markets introduce layers of complexity in financial structuring, impact how the projects are valued, and set the expectation for the business case. Program leaders encountering projects in deregulated markets should recognize some of these issues for their own programs and engage the appropriate experts in a timely manner.

LEADING THE BUSINESS CASE DEVELOPMENT PROCESS

Given all these complexities, leading the business case development process will require astute commercial and business acumen. We have identified five practices beneficial to developing a business case. These practices may seem trivial, but they are frequently overlooked and undervalued.

1. **Be realistic.** Effective programs are firmly grounded in reality, which includes taking into account uncertainty and unforeseen events. These programs assess threats and constraints accurately and own up to failings in a responsible manner. The incentives to overpromise, to not take reasonable risks due to fear of failing, to not seek timely assistance, or to not cut losses and change course are all natural barriers. Effective programs also verify personnel and vendors' abilities and ensure that projects are assigned to people with the right expertise, to avoid delivery failures and project underperformance later.

2. **Be open and aligned about risk and uncertainty.** Effective programs discuss and invite scrutiny in areas that are questionable. These areas can range from promise and performance expectations of unproven technology to new projects to coping with change. Leaders must set directions on screening genuine concerns and fears from issues that are inconsistent with the goals of the program.

Leaders must also set guidelines for alignment. Alignment does not mean driving 100 percent consensus. Many leaders fall into the trap of "death by alignment workshops" to convert every constituent, which is not a practical goal. Alternatively, a heavy-handed approach without a critical mass of staff "on board" is not effective in a multiyear large-scale transformation journey.

3. **Stay in lockstep with the C-suite and the board.** Often short-term issues arise that run counter to the strategic considerations of the program, for example, a sudden financial situation or an unforeseen event can make a change in course necessary. Business cases must have a built-in mechanism to handle such tradeoffs. For instance, a delay in one program can cascade into delays in other dependent programs, requiring a major overhaul of the program economics and scope catching the executives, boards and investors by surprise. Such surprise situations can be avoided if the program is designed and works to ensure transparency and continuous alignment and communication with key stakeholders.

4. **Ensure consistency with the program vision and the program plan.** Companies often undertake visioning, program planning, and design exercises in a sequential manner, and then conduct their activities in a siloed manner. As a result, visioning and strategic choices become dormant and are not shared by business case developers, leading to a disconnect of the business case with the overall picture. Good programs ensure that the business case is in tune with the company's vision and strategic plans, and they set processes and systems in place to ensure they can initiate reviews when those commitments diverge. This practice allows business cases to provide early signals to the programs on potential variances and to take corrective actions quickly.

5. **Develop a public relation, regulatory, and investor relations strategy.** Many companies wait until deep into their program to engage with external stakeholders, mainly because communicating on uncertain outcomes can be a program risk causing increased scrutiny, delays, and holdups. This concern about risk may go too far in terms of lack of communication, and may actually breed

uncertainty, misinformation, and distrust. As we discussed in the AMI example earlier, the industry's lack of public strategy left it unprepared for customers' concerns and outcry regarding health hazards due to radio interference—a result that set back many deployments considerably. Creating a public relation, regulatory, and investor relations strategy that can anticipate such possibilities and proactively shape public discourse will reduce these risks considerably.

SUMMARY

Traditional business case approaches that applied to predictable programs are not appropriate when applied to the set of projects power companies are undertaking under VUCA conditions. Often management's zeal to obtain accurate numerical estimates is not just impractical, but numbers that are gathered are frequently proven to be wrong. This gap between wish and reality, which is often hard to discern, creates friction and an erosion in trust between managers and estimators. Business cases are future-oriented, yet predicting the future has become immensely more difficult; future situations can only be imagined and projected based on the understanding of the forces of change in play. In a dynamic and uncertain environment, programs should view changing cost and benefits estimates as not a bug but a feature, and savvy program leaders should be able to spot and encourage such changes in their teams. Applying a business narrative is extremely useful because it allows leaders and their companies to examine and validate the plausibility and probability of one or more views of possible futures. Business narratives also allow them to incorporate various political, policy, economic, and technical considerations into the model to drive the numbers and allow a more transparent link to regulations. As noted, energy policy expert William Hogan said:

The more innovative the idea, the less likely it will be adopted under the regulated vertically integrated organization. Regulation works better when the choices are highly standardized and easy to evaluate. But when the product includes many moving parts or involves a great deal of uncertainty, it can be much harder to write down the rules. In these cases, business judgments can be made, and the uncertainty can be priced, with the associated risks and rewards.[7]

Leaders creating business cases for power projects are first required to analyze whether the project and the overall program are a good investment. To do so, the program needs to have a coherent, complete, and believable business narrative that can be converted into numbers. The growth of cash flows and the risks associated need to be formulated into a traditional discounted cash flow calculation. Unless there is a compelling financial or business reason, the discount rate can be set to a weighted average cost of capital, with focus lying on the estimation of the cash flow. Once the company reaches a consensus and has confidence that the program is economically viable, the program can then transform the analysis to a regulatory construct that is relevant and expected in the specific jurisdiction. The regulatory strategy and game plan required for rate recovery must be crafted such that they include public outreach, regulatory communications and negotiations, and other measures necessary to seek alignment and approval. The path to approval is not always linear; however, following these guidelines, the business case can be an effective tool that serves as the basis for investment planning, capital allocation, and execution prioritization—and as a communication document for regulators and policymakers. Finally, while developing and using business cases, it is worth remembering the quote, often attributed to Einstein—"Not everything that can be counted counts and not everything that counts can be counted."

NOTES

1 For the basics of corporate finance, readers should refer to any standard graduate textbook, such as Principles of Corporate Finance, by Richard A. Brealey and Stewart C. Myers.

2 For further reading on real options, see Avinash K. Dixit and Robert S. Pindyck, *Investment Under Uncertainty* (Princeton, NJ: Princeton University Press, 1994).

3 Aswath Damodaran, *Narrative and Numbers: The Value of Stories in Business*, Illustrated edition (Columbia Business School Publishing, January 10, 2017).

4 Stephen Denning, *The Leader's Guide to Storytelling: Mastering the Art and Discipline of Business Narrative, Revised and Updated* (Jossey-Bass, Hardcover – February 17, 2011).

5 Modified accelerated cost recovery system.

6 Balance of hedge is a financial hedge that is a form of contract for differences, with a periodic settlement amount that is the net amount owed in one direction or the other after a "fixed payment" made by the hedge provider is netted against a "floating payment" made by the sponsor entity.

7 Electricity Wholesale Market Design in a Low-Carbon Future, *Harnessing Renewable Energy in Electric Power Systems*, 1st Edition, edited by Boaz Moselle, Jorge Padilla, Richard Schmalensee (Routledge).

CHAPTER EIGHT

Blueprinting Program Architecture

What Are the Requirements for Building a Program Architecture as a Solid Foundation for Execution?

DOI: 10.4324/9781003353997-12

In the traditional physical sense, *architecture* is the "art or practice of designing and constructing buildings." The term *enterprise architecture* was borrowed by IT and became popular in the 1990s with the rise of large, complex software systems that were provided by companies like SAP, Cisco, and Oracle. Now the term extends to other parts of the IT business to mean the conceptual structure and logical organization of a computer-based system. Given the complexities of modern grid programs, it is important to develop a conceptual understanding of the program—how its parts fit together, how the organization of the program, which includes not just the operational technology (OT) systems, but also its business activities, organizational groups and hierarchies, grid systems, and enterprise IT—exists logically. In this book, we call this organizational design the *program architecture*. Architecting the overall system and business processes, and envisioning their future state, is a critical step in program execution. The *process* of developing the architecture and the understanding that the company gathers are as vital as the outcomes which take the form of architectural diagrams and artifacts. In fact, the outcome of the architecture efforts, no matter how well defined or how robust, is subject to change. But that is no reason to take this effort lightly; in fact, the potential change is precisely why having a practical, working architecture is so important. Below are a few reasons why programs should invest in a good architecture:

- Good program architecture gives shape and form to the desired functionalities and grounds everyone on a common outlook and goals. When situations change, a good program architecture provides a baseline to manage and adapt change.
- Good program architecture is tightly linked to program design, implementation, and program management. It provides the basis for the tasks, timeline, and resources that will be needed. It provides insights on sequencing major activities and the risks that are involved.
- Good program architecture also provides a view into the company's efforts and the benefits that will be derived through an activity, process, and systems lens. These perspectives elevate awareness

of investments, timelines, and benefits and help in developing quantified estimates in areas such as returns, risks, and cash flow.

This chapter covers the principles and practices that support program leaders in developing good program architecture.

WHY DO TRADITIONAL APPROACHES TO ARCHITECTURE NEED TO CHANGE?

Many business leaders find business architecture to be little more than documents containing detailed representations and diagrams that do not seem to relate to the strategic decisions that these leaders face. As a result, most leaders are removed from the architecture creation process and leave it to IT to manage. This distancing and practice are understandable given that program architecture historically has been limited to large enterprise IT systems. Treated as an execution exercise, architecture development was kept separate from a company's visioning and strategic planning activities. Yet businesses are constantly imperiled by weak or undeveloped architecture. For example, when a large utility began a digital initiative to apply AI/ML to asset lifecycle performance management, it had to spend an enormous amount of time manipulating data that resided in multiple enterprise systems. The processes for asset condition assessment, failure analyses, asset health assessment, and investment planning were fragmented and consisted of duplicated activities and analyses. Decisions regarding whether to replace or repair assets were made by different groups and based on different criteria. One executive remarked, "Every time we do anything digital, it is like starting from scratch, and end up as patchwork." Whether these issues manifest as data problems or process breakdowns, they are, at the root, architectural issues. Without a good foundation, the execution of business initiatives becomes an inefficient and underperforming pursuit.

To cope with VUCA conditions, companies must be agile and able to launch business initiatives with speed. Without a solid architecture as the foundation for operations, agile execution is practically impossible. Companies that do not invest in a good architecture are plagued with recurring challenges in their day-to-day operations. If different functions of the organization come up with different values for the same parameter or different outage restoration times, if several hours are spent reconciling data, if management time is spent more in arbitration or on who should own certain activities rather than in making decisions and taking actions, it's clear that the company has an architecture problem. This condition is particularly important, given that digitalization is no longer an IT capability, but a strategic capability that covers all aspects of the business—from cost controls to new growth creation.

Many companies have established the role of a chief digital officer, reflecting the strategic importance of integrating digital technology across the business. The architecture that is future-oriented will have to address not only back-office integration of data and processes (the focus of traditional enterprise IT), but also core business functions of power generation, transmission, and delivery. As the number of applications grows, the need for a common, trusted source of data is necessary. A good architecture provides greater transparency to enforce decision rights on data access, exchange, and usage to mitigate security risks. As the industry transitions to support decentralized business models with DER, and with increased demand for digital channels like text, chatbots, and voice-assisted devices for customers, the alignment of operations and business needs becomes even more complex. The traditional approach of treating each initiative as a stand-alone, piecemeal solution is no longer sustainable.

Without a comprehensive architectural frame, very soon the company will have a patchwork of different technologies wired with makeshift linkages that are prone to failure and expensive to maintain and extend. Companies that have gone down this route have found themselves dealing with exponentially growing complexity, inflexibility in making changes, and a lack of robustness in operations. In contrast, companies that have invested in solid foundational architecture are able to manage change and respond to market and regulatory needs faster and with greater effectiveness. They turn out to be more resilient to events and quick to pivot and adapt—exhibiting the agility that is necessary under VUCA conditions.

WHY IS ARCHITECTURE A LEADERSHIP COMMITMENT?

Program architecture is often considered to be academic—loaded with reference models, frameworks, and documentation—which is another reason business leaders tend to defer program architecture decisions to IT and stay at arm's length. Though not ideal, in the past, business leaders in power companies adopted such a stance and managed their companies without severe impacts. But in the current environment, business leaders must engage with business architecture. Leaders must be able to choose and must be accountable for the right architecture choice for their business, and they must lead implementation of the architecture.

Major architectural shifts are driven by the development of cloud-based technologies, the internet of things, and data analytics. These technologies have instigated unprecedented architectural options and innovative ownership structures. Initially, these options create cost-reduction options by shifting to public computing, automation, and storage infrastructures. Data and analytics are also enabling new revenue streams and business models centered on customer choice and experiences. No company can ignore these changes; the value proposition is too compelling. Even accounting codes are being updated to remove accounting distortions and barriers. In the US, for instance, the Financial Accounting Standards Board[1] has issued guidelines to alleviate some of the concerns using SaaS (software as a service) and pay-per-use models that in legacy regulations had the disadvantage of being classified as "operations and maintenance" and thus hurt regulated returns.

Although the company's leadership is largely responsible for crafting the company's strategic choices, it is the organization's ability to develop

a purpose-built business architecture that creates the foundation
for executing the strategic choices. Without a commitment to the
architecture, the execution will fall apart. It is through the business
architecture that fundamental decisions regarding how the company will
operate are realized. To get the architecture off to the right start, leaders
must drive clarity in their organization in two major areas:

- Balancing fragmentation and integration of major business
  processes.
- Determining the level of standardization of data and processes.

These two decisions are closely linked to the organizational structure and
governance decisions regarding lines of communication, decentralization
versus centralization, and level of formalization suitable for the
modernization program. In the legacy world, where changes were modest,
business architects with a good grounding on operations were able to
make these decisions. But when the fundamental organization structure
is changing, including new requirements in data management, customer
experience, and innovation, leaders must be at the forefront of strategic
decisions. These decisions will drive high-impact choices on make versus
buy, monolithic systems versus microservices, centralized data warehouses,
such as data lakes versus decentralized and decoupled data, high versus low
latency, real-time processing versus queueing and batch processing, and
rigid design versus flexible. The uncertainty of the operating environment
calls for flexibility and agility, and these characteristics are needed in the
business architecture as well. Employing these characteristics requires
examining new modes of flexible data structures using *extensible schemas*
that began with the use of XML over traditional tight definitions of schemas
requiring a hard tradeoff of one over the other.

Developing an architecture can be a daunting task, not just because of
the technical complexities involved, but also because of the magnitude
of organizational and process changes necessary. Without leadership
commitment, such changes are practically unrealizable. Foundational
changes are major undertakings and require multiple stakeholders,
collaboration, and shared understandings of what needs to be done.

Companies that have ventured into making these changes using the traditional IT-centric approach have encountered numerous challenges. Hence, despite so many thought leaders in the industry championing for the convergence of IT/OT for a while, the efforts in the industry have seen limited success. When business leaders drive architectural decisions, fundamental shifts in the approach occur as well. Under uncertainty, business architects cannot work with a static set of requirements, or with complete information. Rather, businesses have to work with a set of beliefs and hypotheses (see Chapter 3 for more details), an approach different from the traditional linear, deductive practice in which a chain of facts leads to a design decision. Many seasoned architects only accustomed to this approach find it hard to take a decision, with limited information.

Many utilities, especially in Europe, Latin America, and Asia, preferred turnkey systems from vendors that provided integrated systems with pre-built modules that covered many applications for the enterprise. The increased modularity and ability to use application programming interfaces to interact and exchange data created the possibility of a "best-of-breed" approach with shorter time cycles. The approach to development has naturally aligned to Agile and DevOps[2] sprinting versus the traditional phased, sequential, and linear waterfall approach linked with longer deployment cycles.

All the changes stemming from the wide range of available options and technologies add to the complexity of the architecture decision-making. Program leaders may have to take a completely different approach to architecture development; certainly, they need to ensure that the program architects work as a cohesive team, drawing on subject matter expertise as necessary to set up the blueprint for the business.

GUIDING PRINCIPLES (FOR LEADERS)

Creating the architecture blueprint is an exercise in convergence. It is a convergence of the physical with the virtual, the legacy with the needs of the future, the IT and OT, business with IT, and the company with external entities that are part of the end-to-end process and data exchange. It is no longer a project but a journey that requires not just the architecture to be flexible and adaptable but also allows for continuous development, curation, and modification. Success will depend on navigating a large number of options and choices, which can be overwhelming. Below are a set of guiding principles that will increase leaders' chances of success in developing an appropriate business architecture:

- **Tailor the program as a whole.** Drawing on systemic thinking,[3] the architecture must address the "program" as a whole. Individual parts or elements—whether a software, a device, or a group of people (teams or organizational units)—perform an activity or a function, and it is important to think of these parts/elements in the context of and in relationship to the overall program. How they interact, communicate, coordinate, interoperate, and work as an integral part of the program is as important if not more when considering the program holistically. Just as putting a Ferrari engine in a Toyota Prius is not effective, focusing on optimizing one function in the overall program will not ensure the functioning of the overall program. All components must be harmonious, and the architecture must be tailored to the program goals and the vision of the enterprise.
- **Fit with the operating context.** The program architecture must be rooted in a solid understanding of the operating context—not just the current one but also the future one—including talent, skill development, and organizational culture. In addition, it must be prepared to implement the changes necessary to fit the organization to future operations. Many technology projects do this preparation, but often at the back end of an enterprise deployment. The degree of change associated with grid modernization programs can be so dramatic that traditional ways of handling behavior, mindset, and capabilities may not work.

- **Purpose-build the deliverables.** Architecture deliverables are not an end in and of themselves; they are a means to make program execution effective. Large programs that are often implemented in silos are prone to misalignment. For instance, if business architecture development is broken into systems, functional, informational, and organizational representations, developed by different groups of people, then the onus lies on the architecture team to ensure that the overall purpose of the architecture, combining all the representations are met. Usability of these deliverables in downstream program execution activities—be it detailed design, implementation, testing, and operation must be factored into the architecture deliverables. An additional issue is that project updates often happen in isolation without a process to back-feed the changes into the original architecture. Without proper re-evaluation and change control to the architecture, major deviations and inconsistencies can occur. If other dependent projects, unaware of the changes, continue a path that is not in line with the updated program architecture, major rework, delays, cost increases, scope creep, and operational issues may arise. The architecture blueprint is undermined and once it ceases to be practical or usable, loses its pivotal position in the program.
- **Develop a common vocabulary.** Within the architecture, terms and definitions are often used loosely, and they often lack precision in relation to day-to-day activities. A common vocabulary that defines the concepts and relationships for all the program stakeholders is necessary to reduce ambiguity and the risk of errors. Vocabularies can go beyond a simple glossary of terms; they can include semantic definitions that can be used in architecture classifications and that define a particular application, entity, or piece of equipment—or a system's associations, relationships, and constraints.
- **Account for legacy requirements.** As utilities have been building and operating systems for many decades, and given business continuity requirements, programs must be designed and implemented to fit within legacy systems, processes, data, and

governance. The boundaries between the legacy environment and new systems and business processes must be delineated. Architecture must be developed to account for the transition of legacy systems and processes, and it must be developed within the acceptable limits of disruptions. For example, if an asset upgrade program is intended to replace existing analog assets with digital capabilities (e.g., capacitors with communication and computational modules), then the entire system architecture must account for the deployment sequence for capacitors, for communication networks, and for integration with the supervisory control and data acquisition system.

- **Demarcate boundaries.** The overall business and technical architecture need to identify and demarcate the system and organizational boundaries of the program. This is done by using a process of abstraction to classify and create groups that then can be designed, implemented, and tested in a practical manner. During implementation, groups are often treated as individual units but as long as the interfaces and interactions between the groups are defined in sufficient detail, this approach can ensure that the overall system will work. The same goes for physical and logical boundaries. Physical boundaries delineate physical assets, hardware and software systems, discrete manual and automated processes, and software and network components, as well as facilities, service centers, substations, control centers, and other spatial locations. Logical boundaries define classifications, bundling, and packaging of physical components into logical groups. The distinction between physical and logical boundaries is at the foundation of architecture development. For instance, a substation may mean a logical grouping of several physical assets and systems, but as a facility, it may be treated as a physical location.

- **Align program design and management.** Program design is informed by the insights gathered during architecture development. Based on these learnings, it is critical that the original program plan is expanded and adjusted—on timeline, implementation sequence of projects, and stakeholder engagement. Important dependencies between program, projects, architecture, budget, and benefits must be duly harmonized. This step is often overlooked or lacks the necessary rigor, most commonly in the quest for accelerating execution. Architectural guidance represents not just the activities that will be conducted, but also how they will be executed in steps. If a mobile work management system is being deployed, and another project on distribution assets replacement assumes that it will be using the new work management system to schedule and dispatch the crews for the asset deployments, then the program design has to link these deployments. Given that programs can change due to a variety of factors, architectures need to be flexible and scalable. For instance, if new reliability regulations require when certain classes of protection and control devices in the field have to have the latest software security updates, and if the architecture did not foresee the need for remote access from a central location, then physically updating all the protection equipment locally becomes an expensive and inefficient task.

- **Construct for robustness and security.** Given the uncertainty in operating environments, the business architecture needs to clarify the limits of robustness and resilience of the program as a part of its design. This not only will serve as a baseline but also will guide the program on future enhancements and tradeoffs. The rapid growth in adoption of cloud-based solutions, data analytics, and digital platforms is constantly redefining cost structures and business drivers that impact top-level reliability and financial KPIs. Robustness and security are tightly linked. Architecting the physical and logical structures to detect anomalies and to isolate a compromised area quickly from the broader system is critical when dealing with complex integrated systems. A complicating factor is that many legacy systems are too intractable to establish the level of security that is desired. Therefore, understanding the vulnerabilities and providing insights on security are essential parts of the process when developing the business architecture.

ARCHITECTURE GOVERNANCE

Architecture development consumes resources, usually the most experienced and knowledgeable personnel in the organization, and yet the practical value of the entire exercise is often not fully realized or appreciated. This undervaluation has much to do with poorly commissioned and executed architectural efforts that fall apart during the actual build and execution. To deliver good architecture requires robust architecture governance, meaning the process through which decisions are made. Architecture governance is often an overlooked corner of program governance. As a result, the control over architectural decisions and revisions is conducted in a fragmented manner, with ambiguity regarding who is in charge. Architectural governance must adopt a problem-solving approach for architectural decisions, must optimize the architecture blueprinting efforts, and must avoid suboptimal outcomes and wasting time and budgets for the program.

Architecture governance includes a formal structure with responsibilities and accountabilities assigned to activities. The decisions taken range from specific decisions that are linked to a feature of a system or equipment, to enterprise-wide platform decisions that might guide a large set of interdependent functionalities, future design, and system choices. A further complicating factor is that operational conditions and constraints change over time. An external circumstance such as a pandemic, a new regulation, a blackout, or a financial event may require a change in the program, the architecture, the budget, the timeline, the technology, or some other aspect of the business. Such possibilities should be accounted for in architecture governance

Large cross-cutting strategic programs will typically benefit from a central authority that coordinates the architecture decisions and priorities into the most efficient model. Yet not many companies have opted for this approach. An empowered architecture governance body with the right authority ensures that the architecture development is managed according to the agreed principles and guidelines for the program. High-impact, cross-cutting irreversible decisions will normally require a formal decision taken by senior management. In contrast, a functional decision with limited impact on a specific project can be taken within the project architecture team. Such a disposition creates efficiency and reduces risks as decision rights match the severity or impact of the decisions. Typically, architecture governance has seven responsibilities:

1. **Establish a governance body.** The architecture governance body consists of subject matter experts who make architecture decisions. The body must be fully aligned and committed to the overall program vision, strategy, and direction. Many programs disband the body after the first iteration of the architecture, which is a mistake. As programs mature, new insights are gathered that can call for architectural changes. Critical cross-functional decisions may evolve over time. The presence of a standing body helps in maintaining consistency as the architecture evolves with the program. In multiyear programs, the membership will likely change, but if the process is well established, then the program will benefit from continuity.

2. **Identify and support governance processes.** Identify all the activities related to governance (such as checking, reviewing, resolving design conflicts, approving design, and approving alternates and deviations) and define them clearly. The governance body should ensure a system is in place to capture, codify, and amend the processes and decision rights.

3. **Implement a system of controls.** Establish a system of controls that aligns with the governance activities. Implementing such a system includes the parameters for checks, reviews, and approvals to facilitate effective creation and implementation of the architecture and its evolution over time.

4. **Align architectural governance with program governance.** Align architecture governance with the overall program governance, including the central governance body, possible steering committees, and so on, and seek proactive realignment if a deviation from the program objectives that is triggered by the architecture governance becomes necessary.

5. **Establish clear lines of ownership and accountability.** Ensure that the architecture governance body is aware that they will be ultimately held accountable for all architecture decisions. Well-governed programs formalize ownership lines with charters and mandates throughout the program, so that everyone knows who is in charge.

6. **Balance independence and autonomy.** Enable the ability for the architecture governance body to function in a transparent way, which includes operating with a clear set of criteria and a framework through which decisions are made. The workings and decision-making are also to be devoid of influence from other parts of the organization to ensure that no stakeholder can make fundamental changes to the architecture without a review and approval by the architecture governance body. Transparency of these processes leads to clarity and trust.

7. **Foster credibility, transparency, and trust.** Processes and formalized procedures are necessary, but in our experience, they are not sufficient. The effectiveness of the architecture governance body lies in its credibility and the trust it cultivates in its standing and day-to-day workings, as well as the speed with which it responds to change requests, architecture review requests, and clarification requests. It begins with the credibility of the members who compose the governance body as individuals and as members of a group and their ability to function in a constructive and fair manner. Without trust, governance becomes ceremonious, weak, and ineffective, a recipe for almost-certain failure, given the complexity and interdependencies of the program. Trust in the governance process does not happen by accident. It has to be cultivated with genuine commitment. Self-serving decisions, territorial maneuvers, back-channeling, using clout, and organizational standing to steer decision-making are all behaviors that erode trust and that place the entire program at risk.

SUMMARY

To set up operations for the execution demands of the future, power companies need a solid foundation. Investing in a business architecture that allows for underlying integration of data and processes provides that foundation. Because of the increased use of data across the core operations of power businesses, greater integration across traditional functions is necessary. With digital technology embedded in these functions—meaning increased possibilities of automation of tasks, prediction of events, and flow of insights and knowledge to make decisions—business processes must be standardized for seamless access and execution. Leaders no longer can leave data and information management to IT; it is core to how companies operate, engage customers, and generate revenue. Business architecture must be informed and must align with the strategic decisions that leaders make regarding the operations of the enterprise—specifically, on the level of standardization, integration, centralization, and formalization of their processes. Architecture must also align with what and how data is shared, exchanged, and used. While every architecture has to be tailored to the strategic positioning of the company, leaders can apply a few guiding principles in their organizations to make the architecture development process effective. In addition, leaders must ensure adequate governance, such that their organizations have a shared understanding of the architecture, and ad hoc or piecemeal approaches are avoided.

NOTES

1 "Moving to the Cloud? Business Considerations for the New Cloud Computing Accounting Standard," *PwC*, February 2019, HYPERLINK "about:blank"https://www.pwc.com/us/en/services/audit-assurance/ accounting-advisory/cloud-computing.html.

2 Agile is an iterative approach that focuses on collaboration, customer feedback, and small rapid releases. DevOps is a practice of bringing development and operations teams together.

3 Please refer to Peter Senge, *The Fifth Discipline: The Art & Practice of the Learning Organization* (New York: Doubleday, 1990), or Russe L. Ackoff, *Systems Thinking for Curious Managers: With 40 New Management F-Laws* (Triarchy Press Ltd, March 18, 2010).

CHAPTER NINE

Program Design and Management

What Are the Requirements for Program Design and Management?

DOI: 10.4324/9781003353997-13

Power industry's future depends on companies meeting the challenges of a VUCA world, which for most businesses will require effective implementation of a change or modernization program. The success of a program depends a great deal on program management. Program management is defined as:

> *the selection and coordinated planning of a portfolio of projects so as to achieve a set of defined business objectives, and the efficient execution of these projects within a controlled environment such that they realize maximum benefit for the resulting business operations.*[1]

Programs are similar to projects in that they have objectives, budgets, time lines, resources, and performance indicators associated with them; however, in the context of this book, a program is a portfolio of projects. We do not draw a distinction in the capabilities of a program manager versus a project manager. When multiple projects are involved, each project with its own lifecycle, there is a need for a coordinated approach so that the overall set of systemic changes works effectively. In contrast to a company's day-to-day corporate or administrative management functions, program management focuses on a strategic initiative, breaking it into a portfolio of individual projects, each with its own scope, measures, and dedicated resources and budgets that require their own tracking and adjusting. Thus, rather than focusing on the overall performance of the enterprise, programs are designed for specific undertakings. If executed properly, the program keeps people in the organization abreast of its progress and other developments—and ultimately delivers on its objectives. Although project or team leaders oversee the day-to-day execution of the program, business leaders are responsible for the composition of the program portfolio, for the organizational structure of the program, and ultimately, for the outcomes.

WHY DO SO MANY PROGRAMS UNDERPERFORM?

Despite a rich body of practical experience and well-developed academic literature, a significant number of programs underperform and fail to deliver expected outcomes. Many leaders cast these program mishaps as reflecting a generic cultural problem of the organization and point to factors such as risk aversion, resistance to change, and lack of ownership and accountability. Post-mortems of underperforming programs—regardless of regions or jurisdictions—reveal different reasons for failure, however. If these root causes are not acknowledged, then remedies are not explored or implemented. In our practice, the following are the most common reasons modernization programs fail:

- Lack of understanding or denial of complexities,
- Design of programs with projects in silos, leaving gaps in coordination and integration,
- Underestimation of the role of the program owner and the skills required,
- Overreliance on third parties (e.g., consultants and contractors),
- Inability to adapt to changing roles and competencies as the project progresses,
- Limitations in resource deployment,
- Weak governance, lack of timely reviews, and lack of course corrections,
- Lack of clarity regarding the roles of project teams, particularly in cross-functional areas,
- Misalignment between the project and the organization's day-to-day business activities,
- Mismanagement of the way changes are managed with contractors, and
- Misunderstanding of the regulatory implications related to the project.

Although this list is not exhaustive, it illustrates the fact that if a program is launched without adequate structure and design, organization, and leadership, it is more likely to fail. The good news is that with up-front attention to program management, power industry leaders can increase their chances of success in implementing their programs.

Designing the program means constructing a portfolio of projects that is supported by management processes to ensure consistency, repeatability, and performance tracking. Design decisions must be based on a range of parameters: size, complexity, organizational needs, costs and schedule constraints, measures, incentives, and objectives. The projects must be constructed to include the required activities and tasks along with appropriate feedback and control loops to govern their outcomes. Design decisions often involve working with concepts to balance options and trade-offs. Sometimes these decisions appear theoretical and removed from reality, consequently, as the urge to work with tangible activities grows, many design decisions remain unresolved. As activities pick up, if the programs are lucky, with the benefit of new learning, unresolved decisions may become clear, which is great, but often they don't. In such situations, the project can suffer major setbacks resulting in rework and waste, causing cost and schedule overruns.

Another component of program design that often lacks rigor is the program organization itself. As computer scientist Melvin Conway wrote, which is often cited as Conway's law—"Any organization that designs a system will produce a design whose structure is a copy of the organization's communication structure."[2] Communication lines and the hierarchies necessary during the program implementation and operational stage[3] will guide the structure. Simply stated, if more frequent and instant communication across functions is needed, then fewer hierarchies are needed and more cross-functional teams should be engaged; conversely, a strong hierarchy is more effective if tight control of what is exchanged is required. Given the wide variety of projects in most modernization program portfolios, no one formula for the program's organization design will apply to all projects, and a project-by-project assessment is necessary. Like most design efforts, organizational design is an iterative process that requires ongoing changes to the project structure to match needs.

Failed programs are highly correlated to weak program leadership. We distinguish leadership into two distinct roles: (1) the commanding or directing role and (2) the inspiring, motivational, and morale-uplifting role. Both are necessary; however, these two leadership roles are not often found in the same person. An inspiring leader may not have the "director's" skills to direct and ensure top-to-bottom alignment or to reflect, strategize, find creative ways to solve problems in a crisis, or give clear instructions. Senior business leaders often make the mistake of appointing a program leader strong in just one of these roles.

In this chapter, we discuss program design (including our construct of project archetypes) and purpose-built program organization, and more deeply examine leadership and organizational roles.

DRIVERS OF SUCCESSFUL PROGRAM DESIGN

In a VUCA world, a broad array of initiatives and projects will make up the modernization program of a business in the power industry. Projects in a portfolio vary in scope, risks, complexity, and impact; a one-size-fits approach does not work. Projects must be particularly careful in defaulting to the waterfall approach which follows a sequential, linear process with discrete phases, and with no phase beginning until the prior phase is complete. If uncertainty is high, programs will typically invest in experimenting and learning, which includes proof of concepts or pilots. Where speed and agility are needed, program design will include techniques and methodologies such as those of DevOps and Agile software and variants of the design-build-deploy-test processes.[4] In certain complex user-centric situations, such as customer experience and next-generation control systems, design thinking has been found to be useful. Figure 9.1 shows a range of common program design models.

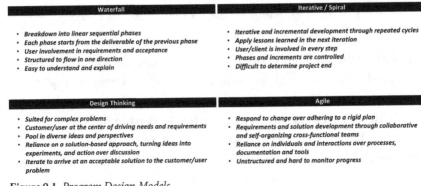

Figure 9.1 Program Design Models

In a VUCA world, there is a broad array of initiatives and projects that are part of the modernization program. Projects in a portfolio vary in scope, risks, complexity, and impact and one-size-fits all does not work. If uncertainty is high, programs will typically invest in experimenting and learning, which includes "proof of concepts" or pilots.

To design a program with such a wide range of underlying characteristics, a rigorous understanding of the project attributes is important. Below are a few examples to illustrate the heterogenous mix of size, complexity, uncertainty, tasks, risks, and technology maturity.

1. *Building enterprise-wide advanced asset management capability:* Requires CapEx investments in upgrading the asset infrastructure. While some old assets have to be replaced, others must be enabled with connectivity, communication to enterprise systems, and advanced analytics that provides insights to optimize asset performance. To realize the value, not only system deployment, but also new processes and upskilling of talent will be necessary.

2. *Development of long duration energy storage or a green hydrogen technology:* Requires a technological breakthrough to break the cost and scaling barrier. Projects are experiments at this stage for learning whether the technology is viable, following which economic and execution feasibility will be determined. The industry requires a combination of policy and public support for economic viability. The energy storage supply chain, and the services for the most part can

be developed independently from other projects that are closer to the core legacy utility operation.

3. *Coupling energy storage with centralized or distributed solar PV:* Harnessing synergies that can positively impact grid operations, but a coupled solar PV and storage project can be commissioned and delivered independently for both the solar PV and the storage parts.

4. *Work productivity improvement:* Depends on the availability and the quality of underlying data bases, such as geospatial data, asset data bases, and capacity utilization metrics.

As mentioned earlier, the main insight here is that a modernization portfolio is likely to consist of a diverse set of project attributes and, hence, program design cannot take a one-size-fits-all approach. Yet at the same time, the program as a whole must be managed such that it can derive synergy benefits and take advantage of portfolio effects. Most traditional entities employ approaches that miss this point in design, and thus they end up managing projects as point solutions or as a group of loosely linked projects.

To address this gap, we introduce a construct called "Project Archetypes." We use the project archetypes as classifiers to group the projects in the portfolio that exhibit similar underlying patterns, that match with familiar projects. As listed in the example portfolio in Table 9.1, there are nine project archetypes, automation, computational/analytics/ optimization, facilities construction, system integration, new products and services, business processes, routine upgrades, events/public relations, and investors/regulatory. Each project archetype has a reference design pattern based on the risks, complexities, familiarity, scope and size, etc. We match the specific projects in the program portfolio to these archetypes to select the reference design suitable for the project. The reference design can then be adjusted for any necessary finer points. This approach allows the application of familiar patterns to new projects and reduces the need to develop individual project designs from scratch for every project.

To determine the appropriate project archetype, program leaders must assess the project attributes and identify project objectives, then choose

Table 9.1 *Project Archetypes*

Project Archetypes	Description
Automation	Increased reliance on information and communication technology to sense and control processes with the goal of eliminating human effort. Given the development in computation, cost reduction, connectivity, digital storage, algorithms, and AI/ML, many possibilities emerge, resulting in frequent changes. New functionalities and performance levels create new frontiers and accelerate obsolescence. The project requires upskilling with a combination of specialists and knowledgeable staff, who in turn require appropriate governance and management models.
Computational/ analytics/ optimization	Increasingly a strategic area that uses emerging technological breakthroughs in large data collection, integration, and analytics to draw insights. Requirements, scope, and timing are hard to know beforehand. Results are uncertain. Highly specialized skills, which are in high demand in the marketplace, are required.
Facilities construction	Routine design built with predictable requirements that are informed largely by historical data. High degree of familiarity resides in the business and the industry, so targets can be set with a high degree of accuracy and activities can be controlled to those targets.
System integration	Involves integrating various components—such as databases, subsystems, and applications—into a cohesive system that delivers a function or process. Multiple stakeholders—vendors, contractors, and subject matter experts (SMEs)—require coordination and control that are not just top down but also across different lateral groups within the organization. Interfacing and interoperability are critical determinants of good system integration.
New products and services	Serves the needs of the customers in the target market by providing value at a price that customers are willing to pay. Such projects require skills for understanding market trends and customer needs, managing research, developing innovation, and undertaking deployment and sales.
Business processes	Involves major overhaul of existing business processes including change in activities, workflows, locations, and interfaces. Process changes impact roles and responsibilities, existing organizational boundaries, communication and interaction, and standard operating procedures. Process changes can be driven by new technology, systems, or operational excellence initiatives.
Routine upgrades	Involves scheduled refresh and upgrades of physical assets including core plant and equipment as well as upgrades in enterprise systems and facilities. These upgrades are usually timebound and predictable with a high degree of certainty in the scope, budget, and timeline.

Table 9.1 *Continued*

Project Archetypes	Description
Events/public relations	Includes specific public campaigns, company events, and engagements that have a specific message, agenda, or a cause. Such activities may have a specific audience or a group that may include activists or influencers. There may be a call to action or to providing information and knowledge. Projects must adhere to specific boundaries that are set by laws, regulations, and policies.
Investors/regulatory	Driven by regulatory policies that include compliance requirements and other mandates that create and support new opportunities to serve constituents. The investment rationale and value creation are dependent and coupled with regulatory provisions and changes in regulatory constructs to create a major shift in the project design, deployment, and outcomes.

the archetype that is the best fit for the program, and finally, customize the project design using the archetype as a model:

1. **Assess project attributes.** The first step is to understand the underlying project attributes and create a profile, which can then be matched and categorized into known project archetypes. In this assessment, identifying uncertainties and project risks is particularly important. Eleven risk drivers are useful for a profile assessment of an initiative or a project in the portfolio:

   - *Novel/unproven technology.* To what extent is the project deploying a novel or unproven technology?
   - *Interdependence of technology.* Does technology used for the program or project rely on other technologies that can delay or prohibit successful implementation?
   - *Multiple disciplines.* Does the program or project require support from multiple disciplines to ensure successful implementation?
   - *Multiple influential stakeholders.* Are there multiple influential stakeholders involved in the implementation of the program that can defer, deviate, delay, or stop the program?
   - *Regulatory/policy/politics exposure.* Is the program or project exposed to regulatory frameworks, policies, or politics?

- *Program or project scope uncertainty.* Is the program or project scope unknown or uncertain with regard to scope, objectives, targets, timelines, budget, resources, etc.?
- *Deployment size and scale.* Is the deployment size and scale of the program or project known?
- *Third-party exposure.* Is the program or project exposed to third parties that can exercise influence over the outcome?
- *Performance achievement.* Is the performance to be achieved by the program or project clearly and objectively defined?
- *Interdependence and handoffs.* Does the program or project have any interdependencies or handoffs with other programs or projects?
- *Vendor/supply chain.* Is there a clear vendor base/supply chain for the elements needed in the program or project?

2. **Record consistent baselines. Program and project baselines need to be establsihed in seven areas.**
   - *Objectives.* Articulate the aim and desired outcomes of the program or project with clarity.
   - *Scope.* Describe the problem/needs/ideas that the program or project will address.
   - *Costs.* Estimate the total costs in terms of capital and operating expenses.
   - *Schedule.* Detail the time and duration of delivering the program or project.
   - *Personnel.* Identify the talent, skills, and teams required.
   - *Organization.* Characterize the mindset, behaviors, and culture necessary.
   - *Uncertainties.* Define the areas of unfamiliarity and unknowns.

3. **Match program design to the project archetypes.** Each project or initiative in the program portfolio is matched with the archetype (depending on the risk profiles created from the assessment of its underlying attributes). Determining the program archetype is ultimately a judgment call, and business leaders and senior executives must be aligned with program leaders to ensure a common, organization-wide understanding on the risk profile of the program portfolio.

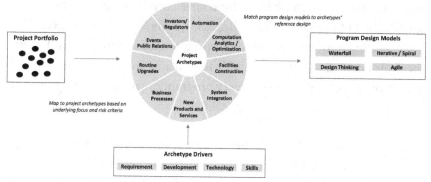

Figure 9.2 Project Design Model Approaches

4. **Create specific design models based on the archetype's reference design model.** Once the projects and initiatives are mapped to the archetypes, suitable program design models can be adopted. The archetypes have proven design models that can be used as references and adapted to specific needs. Figure 9.2 illustrates four broad design model approaches. Senior business leaders will rarely get into the details of program design, but at a minimum, business leaders must ensure that their program leaders are matching a program design model with the risk profile of the project within the context of an uncertain environment.

PURPOSE-BUILT PROGRAM ORGANIZATION

Programs often do not spend as much thought on matching the program to the skills, experience, and temperament of the people who are part of the project organization. In many cases, the governance rules for administration are ad hoc or carry over from the broader enterprise. But program organizations must be purpose-built; composition of the project teams, alignment up the chain and sideways among other groups, channels for communication, methods for collaboration and reporting, and performance feedback loops for course correction must all be specifically fit to program needs. Because no single model will cover all possible situations, it is important for program leaders to learn and adapt the skills necessary to lead and guide their teams and help them adapt and deliver on objectives. To do so, every program organization design should be grounded on three underlying realities:

1. **How large should teams be?** Scaling teams with the increased scope of the program disproportionately increases communication and coordination challenges. Teams typically don't scale with the size of the program; if the size of a team increases along with the program's scope while maintaining the same level of supervision, coordination, and governance, the performance and effectiveness of team members become suboptimal, a situation known as the Ringelmann effect.[5]

2. **How much coordination is needed?** Increased coordination adds overhead with increased administrative tasks. As the number of teams increases, chances of fragmentation and siloed operations increase as well. To avoid inefficiencies, overlap, and redundancies, additional administrative tasks are required. Modernization programs include many different disciplines such as application engineering, user experience, configuration management, data management, communication, analytics, and others, all of which need a certain level of administration.

3. **How closely integrated do teams need to be?** Closer integration allows greater interaction and more holistic treatment but may reduce the speed and efficiency of specialization. When the interactions are complex and closely interlocked, integrated teams are effective, but if the tasks are well defined and can be delegated to specialists, a functionally specialized team can be more effective. Consequently, programs must be thoughtfully crafted to balance the decisions and tasks that need to be centralized with those that could be accomplished in independent groups. It may not always be possible to hit the right balance, so programs need to carefully monitor and adjust the balance based on what works and what does not.

Senior business leaders are ultimately responsible that the organization maps to the program and the cascade into individual projects is aligned with the overall purpose, and that the design criteria are consistently applied to fulfill the program's purpose.

DEFINED ROLES AND RESPONSIBILITIES

Program management consists mainly of coordinating the various functions, teams, experts, and stakeholders, both internal and external that are associated with underlying projects. Program leaders typically create the playbook that describes the coordination, review, and governance processes and mechanisms. A well-proven and transparent structure to manage modernization programs has five layers:

1. Board and company leadership,
2. Program sponsors,
3. Program governance body,
4. Program management office (PMO), and
5. Program managers.

A transparent approach serves most complex programs well, particularly because it clarifies for delivery teams how projects are reviewed, how issues are identified, how risks are discussed, and how problems are addressed. Good program leadership must understand how and when to engage senior leadership and the board. Finally, given that the board and senior business leaders tend to be the program/project sponsors, it is they who approve investments. We discuss each of these five layers further in the following sections.

ROLE OF THE BOARD AND COMPANY LEADERSHIP

The board of directors and the company's leadership are the faces of the company and the main messengers regarding the program to both internal and external stakeholders. Several external factors can govern the board agenda from activist shareholder interventions, societal unrest, and natural disasters to political turmoil. When such situations arise, while large transformational programs are underway, it can be quite disruptive and can consume management time. However, there are certain roles the board and senior leaders must play to ensure continuity and overall success of the program:

- Create awareness of the program for both internal and external stakeholders,
- Communicate the vision and the underlying drivers for the program,
- Allocate resources that reflect commitment to the program,
- Defend the decisions made by the program governance body and the PMO,
- "Walk the talk" to maintain the program's credibility and integrity, and
- Ensure succession planning is consistent with multiyear program commitments.

ROLE OF THE PROGRAM SPONSORS

Program sponsors ensure that the program, with all its projects, delivers the desired business outcomes and results. Program sponsors act as representatives of the organization, and they have the following responsibilities:

- Articulate the overall strategic intent, direction, and business context of the program,
- Remind stakeholders of the bigger picture and the rationale for the program,
- Provide oversight and guidance, as necessary,
- Communicate the overall constraints and boundaries of the program,
- Actively maintain alignment with project teams,
- Serve as strong advocates for the program and seek alignment with internal and external stakeholders,

- Seek the necessary resources to deliver the program: capacity, funding, and priority,
- Serve as the escalation point for matters that are beyond the authority of the PMO, and
- Act as the conduit for communication between the program and external stakeholders.

ROLE OF THE PROGRAM GOVERNANCE BODY

Large programs are cross-functional and impact multiple business and corporate support functions. To ensure alignment on inputs, assumptions, risks, and outcomes, governance bodies are set up for oversight and to facilitate decision-making. Specialized governance bodies include those for thought leadership, architecture, and technical matters. The program governance body typically consists of senior representatives of the internal stakeholders in the program. The responsibilities of the program governance body include:

- Ensure that the company's vision and strategic choices are served by the program,
- Acquire, secure, and allocate resources in time for projects according to their prioritization,
- Be accountable along with the PMO for the successful completion of projects under the program,
- Validate the stage gates[6] between project phases, and ensure that phase transitions meet the established acceptance, modification, or exiting criteria, and
- Resolve conflicts and ensure established guidelines are followed with consensus and arbitration as necessary.

ROLE OF THE PROGRAM MANAGEMENT OFFICE

The core structure that most modernization programs deploy to manage the program and its combination of projects is the program management office. The PMO is a popular construct for its ability to provide dedicated attention and focus to the program, removed from the day-to-day business operations. The PMO conducts several tasks that cut across projects such as planning, vendor selection, risks, architecture validation, finance, and reporting:

- *Program Management*: Responsible for the program design and execution across all workstreams and projects. Reports to and is a member of the program governance body.
- *Integrated Planning, Scheduling, and Tracking*: Manages an integrated time and resource plan across workstreams.
- *Business Architecture:* Manages an integrated business process architecture across the organization based on the appropriate business requirements, including design reviews and process validations.
- *Technology Architecture*: Manages an integrated technology architecture across the organization based on the appropriate technical requirements, including design reviews and technology validation.
- *Financial Control:* Manages the program budget across workstreams, performs cost management and accounting, and performs financial analysis and reporting.
- *Strategic Direction:* Tracks the business case against the strategy, manages the program scope, undertakes research and strategic benchmarking, and leads periodic strategy updates.
- *Change Management and Communication:* Manages program communication with internal and external stakeholders, develops marketing and information campaigns, sets up and manages the change management process.
- *Vendor Management*: Handles vendor selection and screening, manages vendor relationships, and coordinates centralized procurement.
- *Risk Management*: Manages risk profiles, oversees risk-mitigation processes and execution, and issues reporting and mitigation activities.

ROLE OF THE PROGRAM MANAGERS

The program manager role is the key leadership role for the coherent, integrated, and coordinated implementation of the modernization program. Under the day-to-day leadership of the program manager, the projects are defined, designed, and implemented such that the vision is realized, expected benefits are obtained within the timeframe envisioned, and costs are managed according to budget. The program manager role requires several important traits and skills that should be validated and confirmed by senior business leaders, including:

- Possess detailed project-level understanding to provide direction to the teams,
- Know how to identify and mitigate risks,
- Have great people skills and act as a liaison between the program and company leadership, governance committee, and the board,
- Understand and enforce governance rights,
- Know how to implement the overall program master plan, and
- Manage program scope, timeline, and budget.

CREATING A CONDUCIVE CONTEXT

Seasoned business leaders know that program delivery and management do not always follow detailed project plans. Good program managers display a relentless focus on reality and ground their project decisions on facts; by contrast, when projects deviate from plan, many program managers are reluctant to accept reality and may employ wishful thinking, often because senior leaders don't want to hear or share bad news to avoid the appearance of underperforming and/or punitive repercussions. Such behavior creates an environment that chronically underestimates risks. Attention and remediation come only when operating systems are rapidly failing and falling apart. To avoid such situations, good leaders make it a priority to create an environment (including processes, mechanisms, and practices) that is conducive to collecting facts, drawing insights, and making course corrections quickly. If matters need to be escalated to senior leaders, they should be communicated transparently and not stifled by fears of retribution. Given the pervasiveness of uncongenial program environments in the industry, expecting a program leader to be able to create a perfectly conducive working context may often be dismissed as a Panglossian view. But not only are the stakes high if a program fails, but the rewards are also worthwhile if leaders actively prioritize and promote behaviors that support a positive work environment and discourage those behaviors that can create a toxic work environment.

Leaders know that six operational enablers (Table 9.2) must work in cohesion to enable a robust program management capability. Senior leaders must ensure that program leaders do what is necessary to develop and leverage these enablers to create a working environment of transparency and fact-based analyses and decision-making. The cooperating and competitive behaviors of the teams must be in balance and be consistent with the needs of the program. There should be openness among teams, and there should not be perverse incentives for team members to overestimate their abilities to deliver or to not stretch themselves because they fear failure. The stakes for modernization programs are particularly high, but applying the "no-mistake, no-failure" benchmark for legacy routine operations on a first-of-a-kind program is also unrealistic. Particular attention should be given to the growing number of first-of-a-kind projects, which rarely have robust best practices that can provide a template for design and delivery.

How people behave in organizations are largely driven by how incentives are set up and disbursed. When teams are not clear how their measures, monetary incentives, and career progression will work, members may experience anxiety and exhibit unwanted behaviors, such as lack of collaboration and information sharing. Programs may need to develop their own performance management systems—with metrics and KPIs that match project needs. Teams may need incentives different from those of the legacy businesses as well as different criteria for promotion. Program

Table 9.2 *Program Management Capability Operational Enablers*

Operational Enablers	Description
Governance and operating model	Organization structure
	Roles and responsibilities
	Decision rights
Management and control system	Program management
	Risk management
	Configuration management
	Customer management
	Information flows
	Measurements and incentives
	Coordination mechanisms
Execution	Engineering and design
	Supply chain management
	Project execution
	Construction
Expertise	Talent management
	Domain expertise
Enabling	Information technology
	Infrastructure
	Process management
	Compliance
Culture	Risks
	Innovation
	Engagement
	Focus

leaders often don't have the authority to change incentives, especially in organizations where Human Resources operates at arms-length and programs are not able to direct or influence incentive structures. In these cases, it is important for senior leaders to align incentives to the behaviors and to the projects. Such situations compound the difficulty in driving team behaviors with limited options at the program manager's disposal, requiring even greater diligence in finding tools to drive behavior. There are several operational and leadership components that contribute to the quality of the program environment, which has an enormous impact on the program outcomes. The onus lies on program leadership to cultivate those practices in their program organizations.

SUMMARY

Given the VUCA nature of today's environment, the power industry's modernization efforts will require a broad range of approaches for program design and program management. Particular attention should be given to the growing number of first-of-a-kind projects, which rarely have robust best practices that can provide a template for design and delivery. In the past, companies had the option of either (1) waiting for the uncertainty to clear or (2) cautiously following a first mover as a means to de-risk and hedge their bets. In the current environment, those options are not always available, as technology refresh cycles have become shorter, and the pace of change has accelerated. Program management needs to adapt and account for planned obsolescence. In this chapter, we presented an approach for grouping a program's projects by "project archetype" and fitting them with different program design options, from waterfall to design thinking, depending on the nature of the project. The program organization is structured to make it easy for necessary collaboration and communication. Team members and the broader company leadership have defined roles and responsibilities, and they share an understanding of governance processes though the success of realizing a company's goals resides in the activities conducted by the working-level teams. Although senior business leaders are not involved in the daily details of program management, they play a critical role. They must ensure the program design matches the program portfolio to deliver on the company's modernization objectives, and they must provide a conducive operating context—which includes clarity on the shared purpose, expectations, measures of success, and incentives that align with good behaviors—a priority.

NOTES

1 Max R. Wideman, *Wideman Comparative Glossary of Project Management Terms v. 5.5*, http://www.maxwideman.com/pmglossary/PMG_P11. htm#Program%20Management. Accessed May 2021.

2 Melvin E. Conway, "How Do Committees Invent?" *Datamation* 14, no. 4 (April 1968), 28–31.

3 *Operational stage* means post-implementation, when the project is in commercial operations.

4 For a more detailed understanding of DevOps: Gene Kim, *The DevOps Handbook: How to Create World-Class Agility, Reliability, and Security in Technology Organizations* (IT Revolution Press, 2016).

5 Max Ringelmann, "Recherches sur les moteurs animés: Travail de l'homme" [Research on animate sources of power: The work of man], *Annales de l'Institut National Agronomique*, 2nd series, vol. 12 (1913), 1–40, Available online in French at http://gallica.bnf.fr/ark:/12148/bpt6k54409695.image.f14. langEN.

6 Stage gate or phase gate is a project management technique to break up projects into a series of stages of phases with gates between them as review and decision points whether to modify, continue, or end the program.

INDEX

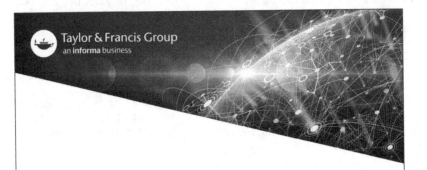

Printed in the United States
by Baker & Taylor Publisher Services